Springer Series in
OPTICAL SCIENCES   133

Founded by H.K.V. Lotsch

Springer Series in
OPTICAL SCIENCES

The Springer Series in Optical Sciences, under the leadership of Editor-in-Chief *William T. Rhodes*, Georgia Institute of Technology, USA, provides an expanding selection of research monographs in all major areas of optics: lasers and quantum optics, ultrafast phenomena, opti-cal spectroscopy techniques, optoelectronics, quantum information, information optics, applied laser technology, industrial applications, and other topics of contemporary interest.

With this broad coverage of topics, the series is of use to all research scientists and engineers who need up-to-date reference books.

The editors encourage prospective authors to correspond with them in advance of submitting a manuscript. Submission of manuscripts should be made to the Editor-in-Chief or one of the Editors. See also www.springer.com/series/624

David L. Andrews   Zeno Gaburro
(Editors)

Frontiers in Surface Nanophotonics

Principles and Applications

 Springer

David L. Andrews
Nanostructures and Photomolecular
  Systems
School of Chemical Sciences
University of East Anglia
Norwich NR4 7TJ
U. K.
david.andrews@physics.org

Zeno Gaburro
INFM e Dipart. di Fisica, Universita'
  di Trento,
via Sommarive 14, I-38050 Povo (TN),
Italy
gaburro@science.unitn.it

ISBN-13: 978-1-4419-2377-6   e-ISBN-13: 978-0-387-48951-3

Printed on acid-free paper.

9 8 7 6 5 4 3 2 1

springer.com

Preface

Surfaces and interfaces of all kinds determine our primary sensory experience of the world we inhabit, and our perception of solidity is very largely based on an interpretation of surface data. In the mind of the scientist, a process of abstraction has led to the familiar modeling of surfaces as two-dimensional boundaries between different parts of otherwise three-dimensional systems. Indeed, such 'ideal' interfaces satisfactorily engage with simple macroscopic models to describe mechanical, acoustical or optical phenomena as they appear to human perception. Such connections work because, on the length scale of a meter (and a few orders of magnitude above or below), the various dynamical responses of material systems are dominated by the bulk properties of their components.

The reliability of this perspective has been significantly undermined by the advent of nanotechnology. As the reader will find in most introductory presentations, the prefix *nano* derives from the Greek for 'dwarf', and the new usage follows its earlier adoption in quantitative unit terminology to denote a factor 10^{-9}: indeed, nanotechnology deals with systems where the scale of typical lengths is from one up to some tens of nanometers ($1nm = 10^{-9}$meter). To see why this makes a difference to the relationship between surface and bulk properties, it is instructive to reflect on the deeper physical meaning of such a length scale, which corresponds to a typical length of a few atomic bonds.[1] This means that in *nanomaterials* – materials whose dimensions are confined to the nanometer scale – atoms within the 'bulk' are never more than a few bond lengths away from an interface of some kind. Consequently it is no longer possible to consider any intrinsic property of such a material as independent from its boundary properties. Moreover, in some sense we could say that in shifting focus from the macro- to the nano-world, we find materials shifting from primarily bulk-like to surface-like attributes, and it is their surface rather than bulk properties that determine system behavior.

There is a whole realm of fascinating phenomena associated with surfaces; although they afford stimulating opportunities to creative minds, their understanding it is by no means an easy endeavor: "God created all matter—but the

---

[1] The bond length scale is more appropriately located in the Ångstrom range ($1\text{Å} = 0.1nm$).

surfaces are the work of the Devil" was Wolfgang Pauli's reported view. In this book, we touch on a number of recently developed concepts relating to the nanoscale optical and photonic properties of surfaces, without any attempt at completeness. Some of the ideas are still speculative, as in the case of Chapter 1 which addresses the interaction of light with moving interfaces. Such a concept is particularly intriguing in connection with photonic structures where light can be significantly slowed down – and where it is even possible to conceive superluminal effects. Chapter 2 steers the subject matter to topics of already proven practical application, the use of evanescent optical waves for sensing applications. Of course, the device implementation of any form of nanoscale response is ultimately subject to the development of efficient and inexpensive fabrication techniques, to achieve the necessary nanostructures. In Chapter 3, in the context of potential applications in biology, attention focuses on a case in point, porous silicon, which has captured much scientific attention over the last fifteen years and now seems well placed for significant commercial exploitation. A survey of integrated optoelectronic circuits in Chapter 4 puts us back in the context of one of the most relevant field of applications, where optics and nanotechnology are strongly pushing research and development. Metallic nanostructures are also of major interest, especially for phenomena related to surface plasmons (mobile charge oscillations at metallic-dielectric interfaces): Chapter 5 reviews the properties and applications of gold quantum dots. Finally, Chapter 6 concludes with another topic at the adventurous forefront of the subject, concerning the highly unusual optical properties of surfaces fabricated to develop a chiral topography.

We belive that the interplay of surface properties and optics will, in the years to come, more and more emerge as a broad but distinctive field of photonics, one that is rich both in its fundamental interest and its potential for creating major applications. We hope, with this book, to stimulate that interest and the creativity in our readers.

Zeno Gaburro, Trento
David L. Andrews, Norwich

February 2007

Contents

Contributors

Zeno Gaburro INFM e Dipart. di Fisica, Universita'di Trento, via Sommarive 14, I-38050 Povo (TN), Italy

Silvia Mittler Department of Physics and Astronomy, The University of Western Ontario London, Ontario N6A 3K7, Canada

Bernhard Menges Max-Planck-Institut für Polymerforschung, Ackermann Weg 10, 55128 Mainz, Germany

Huimin Ouyang Department of Electrical and Computer Engineering and Center for Future Health, University of Rochester, Rochester NY 14627, USA

Marie Archer Department of Electrical and Computer Engineering and Center for Future Health and Department of Biomedical Engineering, University of Rochester, Rochester NY 14627, USA
and
Permanent address: U.S. Naval Research Laboratory, Center for Biomolecular Science and Engineering, Washington D.C. 20375

Philippe M. Fauchet Department of Electrical and Computer Engineering and Center for Future Health and Department of Biomedical Engineering, University of Rochester, Rochester NY 14627, USA

David V. Plant Department of Electrical and Computer Engineering, McGill University

Luca Prodi Department of Chemistry "G. Ciamician", University of Bologna via Selmi 2, 40126 Bologna Italy

Gionata Battistini Department of Chemistry "G. Ciamician", University of Bologna via Selmi 2, 40126 Bologna Italy

Luisa Stella Dolci Department of Chemistry "G. Ciamician", University of Bologna via Selmi 2, 40126 Bologna Italy

Marco Montalti Department of Chemistry "G. Ciamician", University of Bologna via Selmi 2, 40126 Bologna Italy

Nelsi Zaccheroni Department of Chemistry "G. Ciamician", University of Bologna via Selmi 2, 40126 Bologna Italy

Fei Wang Micron Technology, Inc. 8000 S. Federal Way. P. O. Box 6 Boise, ID 83707-0006, USA

1

Moving Dielectric Interfaces as Photonic Wavelength Converters

Zeno Gaburro

University of Trento, Italy
gaburro@science.unitn.it

1.1 Introduction

A distinctive characteristic of optical communications is the ability of parallel transmission and signal processing. To fully exploit this feature, one needs not only tunable light sources and fast detectors, but also complex devices such as add-drop filters, routers, and wavelength converters [1].

Several approaches are possible regarding wavelength conversion [2, 3]. They can be grouped into two classes, as either all-optical or electro-optical techniques. In the former case, optical signals are directly processed and frequency-shifted in the optical domain. All-optical means include for example cross-gain and cross-phase modulation, and four-wave mixing in semiconductor optical amplifiers [3]. The other option is electro-optical techniques, for which signals must be first transformed into electrical signals, processed, and then back transformed into optical signals. The advantage of the second approach stems from the fact that signal processing is conducted inside the electronic domain, which relies on established technologies. The disadvantage is that it clearly requires additional overhead for the optical–electrical–optical (OEO) conversions. Moreover, electrical effects have generally slower time constants than optical effects and thus lower bandwidth available. Technology of electro-optical converters is complex, suggesting significant power consumption and large number of components and costs when a high bit rate is required [2].

It is well known that reflection and refraction by moving interfaces lead to Doppler shift of frequency of the incident waves, and therefore can be proposed as an all-optical solution for wavelength conversion [4, 5]. Compared to other physical effects — including for example the Raman effect — the Doppler shift provides a way to *continuously* tune photon energies. In this work we show that reflection and refraction by dielectric interfaces in inertial frames can be directly interpreted as *photonic* phenomena — in the sense that they act on photons — even within a fully classical description. The photon balance at the interface is indeed both conserved and invariant under any inertial motion of the interface. Conservation of photon balance means that if we respectively label as N_i, $N_r = \alpha_r N_i$ and $N_t = \alpha_t N_i$ the number of incident, reflected and refracted photons per unit time per unit area (where $0 \leq \alpha_r, \alpha_t \leq 1$), then $\alpha_r + \alpha_t = 1$, in any inertial frame. Invariance means that α_r and α_t do not depend on the inertial frame. As we show, the values of α_r and α_t are the well-known power reflection and refraction coefficients, as calculated in the frame of reference at rest with the interface. Conservation and invariance of photon fluxes stay jointly consistent with the conservation of energy and with Doppler shifts.

An interesting option to obtain an interface moving at high speed is to modulate in space and time the dielectric function of the medium without putting it into real motion. This is possible if the materials exhibit a significant electro-optical effect, and if this effect can be driven by an injected electromagnetic wave — for example, by a microwave, as done in traveling wave modulators. We show that conservation and invariance of photon fluxes hold even in this case, in which there is relative uniform motion between the interface and the media.

1.2 Electromagnetic Waves in Inertial Dielectric Media

Maxwell's equations in the SI system of units are [6, 7]

$$\nabla \cdot \mathbf{B} = 0; \quad \nabla \times \mathbf{E} + \frac{\partial \mathbf{B}}{\partial t} = 0; \quad \nabla \cdot \mathbf{D} = \varrho_f; \quad \nabla \times \mathbf{H} - \frac{\partial \mathbf{D}}{\partial t} = \mathbf{J}_f. \qquad (1.1)$$

where the subscript f refers to *free* charge and current densities. The first two equations above allow the introduction of potentials Φ and $\mathbf{A}$ such that

$$\mathbf{E} = -\nabla \Phi - \frac{\partial \mathbf{A}}{\partial t}, \quad \mathbf{B} = \nabla \times \mathbf{A}. \qquad (1.2)$$

The effect of bound (or polarization) charges is included in the relationships

$$\mathbf{D} = \epsilon_l \mathbf{E}, \qquad \mu_l \mathbf{H} = \mathbf{B}, \qquad (1.3)$$

where ϵ_l and μ_l are, respectively, the dielectric and magnetic permittivity of medium l. All media are here taken as nonabsorptive (ϵ_l and μ_l are purely real), isotropic (the dielectric and magnetic tensors are scalar $\epsilon_l^{ij} = \epsilon_l \delta_{ij}$, $\mu_l^{ij} = \mu_l \delta_{ij}$), and nondispersive ($\epsilon_l(\omega) = \epsilon_l$, $\mu_l(\omega) = \mu_l$). This latter condition implies that the *phase* and *group* velocity of light coincide, and that the medium response is local in space and time.

We will indicate with ϵ_0 and μ_0 the permittivities of vacuum. Equations (1.3), together with

$$\mathbf{J} = \sigma_l \mathbf{E} \qquad (1.4)$$

are known as *constitutive equations*, as they express the role of the material. In the following, the medium of incidence will be labeled as medium 1, and the medium of transmission as medium 2.

Equations (1.1) and (1.2) hold in any inertial frame of reference, whereas Equations (1.3) and (1.4) are valid only in an inertial frame at rest with the media,[1] i.e., they are not covariant [6]. Their general extension to any inertial frame was first derived by Minkowski in 1908 [8]. According to the special theory of relativity, fields $\mathbf{E}$, $\mathbf{B}$, $\mathbf{D}$, and $\mathbf{H}$ can be expressed as four-dimensional tensors $\mathbf{F}$ and $\mathbf{G}$ as

$$\mathbf{F} = \begin{bmatrix} 0 & -E_x & -E_y & -E_z \\ +E_x & 0 & -cB_z & +cB_y \\ +E_y & +cB_z & 0 & -cB_x \\ +E_z & -cB_y & +cB_x & 0 \end{bmatrix},$$

$$\mathbf{G} = \begin{bmatrix} 0 & -cD_x & -cD_y & -cD_z \\ +cD_x & 0 & -H_z & +H_y \\ +cD_y & +H_z & 0 & -H_x \\ +cD_z & -H_y & +H_x & 0 \end{bmatrix}.$$

(1.5)

We label as O the frame of reference of the laboratory, and as O' the frame of reference stationary with the moving media. Physical quantities are labeled accordingly. The relationships between electromagnetic tensors as expressed in the two different frames are given by

$$\mathbf{F} = \mathbf{L}' \cdot \mathbf{F}' \cdot \mathbf{L}'^{T}, \quad \mathbf{G} = \mathbf{L}' \cdot \mathbf{G}' \cdot \mathbf{L}'^{T}, \tag{1.6}$$

where $\mathbf{L}'$ is the Lorentz transformation matrix $O' \to O$ (the direct transformation $O \to O'$ being labeled as $\mathbf{L}$) [9]:

$$\mathbf{L}(\boldsymbol{\beta}) = \begin{bmatrix} \gamma & -\gamma\boldsymbol{\beta}^{T} \\ -\gamma\boldsymbol{\beta} & \mathbf{I} + \dfrac{\gamma^2}{\gamma+1}\boldsymbol{\beta}\boldsymbol{\beta}^{T} \end{bmatrix}, \quad \boldsymbol{\beta} = \begin{bmatrix} v_x/c \\ v_y/c \\ v_z/c \end{bmatrix}, \quad \gamma = \frac{1}{\sqrt{1-\boldsymbol{\beta}^T \cdot \boldsymbol{\beta}}}, \tag{1.7}$$

$$\mathbf{L}'(\boldsymbol{\beta}) = \mathbf{L}(\boldsymbol{\beta})^{-1} = \mathbf{L}(-\boldsymbol{\beta}).$$

In the definitions (1.2), $\mathbf{I}$ is the three-dimensional identity matrix, and $\boldsymbol{\beta}$ is the velocity of frame O' with respect to O, where $c = 1/\sqrt{\epsilon_0\mu_0}$ is the velocity of light in vacuum.

The general constitutive relations in the laboratory frame are found by expressing the fields in O' as functions of the fields in O (by using the direct Lorentz transformations, $\mathbf{L}$), and inserting them into Equations (1.3). The result is

$$\mathbf{D} + \frac{\boldsymbol{\beta}}{c} \times \mathbf{H} = \epsilon_l(\mathbf{E} + c\boldsymbol{\beta} \times \mathbf{B}), \quad \mu_l(\mathbf{H} - c\boldsymbol{\beta} \times \mathbf{D}) = \mathbf{B} - \frac{\boldsymbol{\beta}}{c} \times \mathbf{E}. \tag{1.8}$$

---

[1] Hereafter, we assume that the frame at rest with the media is inertial.

If Equations (1.8) are plugged into Maxwell's equations (1.1), a general tensor wave equation is obtained [10], which is valid in any frame in uniform translation with respect to the frame at rest with the media.

If no free charges nor free currents are present in the dielectric, by choosing the Coulomb gauge ($\nabla \cdot \mathbf{A} = 0$ and $\Phi = 0$) for the potentials, the general wave equation simplifies as [10]

$$\left[\nabla^2 - \frac{1}{c^2} \frac{\partial}{\partial t^2} - (n_l^2 - 1)\gamma^2 \left(\boldsymbol{\beta} \cdot \nabla + \frac{1}{c} \frac{\partial}{\partial t} \right)^2 \right] \Phi = 0,$$

$$\left[\nabla^2 - \frac{1}{c^2} \frac{\partial}{\partial t^2} - (n_l^2 - 1)\gamma^2 \left(\boldsymbol{\beta} \cdot \nabla + \frac{1}{c} \frac{\partial}{\partial t} \right)^2 \right] \mathbf{A} = 0,$$

(1.9)

where we have defined the refractive index as

$$n_l = \sqrt{\frac{\epsilon_l \mu_l}{\epsilon_0 \mu_0}}.$$

(1.10)

The solutions of Equations (1.2) are plane waves. Thus, plane waves are the general solutions of Maxwell's equations in any inertially moving rigid medium, in the absence of free charges and currents. The dispersion relation of a general solution of Equations (1.2) is [10]

$$k^2 - \frac{\omega^2}{c^2} - (n_l^2 - 1)\gamma^2 \left(\frac{\omega}{c} - \boldsymbol{\beta} \cdot \mathbf{k} \right)^2 = 0.$$

(1.11)

In the frame at rest with the medium, Equations (1.2) reduce to the well-known special case

$$\left[\nabla^2 - \frac{n_l^2}{c^2} \frac{\partial}{\partial t^2} \right] \Phi = 0,$$

$$\left[\nabla^2 - \frac{n_l^2}{c^2} \frac{\partial}{\partial t^2} \right] \mathbf{A} = 0,$$

(1.12)

with dispersion

$$k^2 - \frac{\omega^2 n_l^2}{c^2} = 0.$$

(1.13)

The phase velocity of plane waves which are solutions of Equations (1.12) is $v_l = c/n_l$. The medium impedance is defined as [9]

$$\eta_l = \sqrt{\frac{\mu_l}{\epsilon_l}}.$$

(1.14)

In the following, we shall restrict for simplicity to the case where both the light propagation and the interface motion are in the direction normal to the interface.

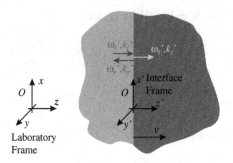

FIGURE 1.1. Frames of reference.

In Figure 1.1, the frame O' — the *frame of the interface* — is at rest with respect to the interface. We set the x' and y' axes of O' to lay in the plane of the interface, and the z' axis to be perpendicular to the interface. The interface is moving at velocity $\mathbf{v} = v\mathbf{u}_z$, with $v > 0$, with respect to the reference frame O.

We start by considering in Section 1.3, the standard case of two rigidly moving media of different permittivities $(\epsilon_{1,2}, \mu_{1,2})$ separated by a flat interface. In this case, which we refer to as the *material interface case*, the two media are at rest in the interface frame.

In Section 1.4, we extend the discussion to the case where the interface is traveling with respect to the media. In this *traveling interface case*, the media are at rest in the laboratory frame O, but their permittivities are modified in time by a traveling electromagnetic wave via the electro-optical effect which creates an effective interface between regions of different permittivities.

1.3 Material Interface Case

For the observer O', the case under study is just the classical case of reflection and refraction by the interface between two media at rest. The phase matching condition at any point of the still interface requires that a single plane wave generates a single reflected and a single refracted plane wave, and that frequency is conserved [6, 7, 11]. We label the frequency of any wave as ω' ($\omega' = \omega'_i = \omega'_r = \omega'_t$). The following relationships hold for the k-vectors and phase velocities:

$$k'_i = \frac{\omega'}{c} n_1, \quad k'_r = -\frac{\omega'}{c} n_1, \quad k'_t = \frac{\omega'}{c} n_2,$$

$$v'_{ph,i} = \frac{c}{n_1}, \quad v'_{ph,r} = -\frac{c}{n_1}, \quad v'_{ph,t} = \frac{c}{n_2},$$

(1.15)

where the subscripts i, r, t refer, respectively, to incident, reflected, and refracted plane wave.

The Poynting vectors $\mathbf{S}' = \mathbf{E}' \times \mathbf{H}'$ in O' are

$$\mathbf{S}'_i = +\frac{E_i'^2}{\eta_1}\mathbf{u}'_z, \quad \mathbf{S}'_r = -\frac{\varrho'^2 E_i'^2}{\eta_1}\mathbf{u}'_z, \quad \mathbf{S}'_t = +\frac{\tau'^2 E_i'^2}{\eta_2}\mathbf{u}'_z,$$

$$S'_i = \frac{E_i'^2}{\eta_1}, \qquad S'_r = -\frac{\varrho'^2 E_i'^2}{\eta_1}, \qquad S'_t = \frac{\tau'^2 E_i'^2}{\eta_2}, \tag{1.16}$$

where we have introduced the symbols ϱ' and τ'

$$\varrho' = \frac{E'_r}{E'_i} = \frac{(\eta_2 - \eta_1)}{(\eta_2 + \eta_1)}, \quad \tau' = \frac{E'_t}{E'_i} = \frac{2\eta_2}{(\eta_2 + \eta_1)}, \tag{1.17}$$

which are, respectively, the reflection and refraction coefficients of the electric fields in the frame O'. They are obtained by applying the boundary condition (equal tangential components in media 1 and 2) to the electric field $\mathbf{E}'$ and to the magnetic field $\mathbf{H}'$.

The electromagnetic energy density $w' = 1/2(\mathbf{E}' \cdot \mathbf{D}' + \mathbf{B}' \cdot \mathbf{H}')$ in O' is, [11] for each wave,

$$w'_i = \frac{E_i'^2}{c\eta_1}, \qquad w'_r = \frac{\varrho'^2 E_i'^2}{c\eta_1}, \qquad w'_t = \frac{\tau'^2 E_i'^2}{c\eta_2}. \tag{1.18}$$

The conservation of energy at the interface can be expressed as

$$S'_i + S'_r - S'_t = 0, \tag{1.19}$$

as obtained by substituting Equations (3.1) into Equations (1.16). In Equation (1.19), S'_i and S'_r are added with positive sign because the z' axis points from medium 1 toward the interface, whereas S_t is added with negative sign because the z' axis points from the interface toward medium 2. Since the Poynting vector and the electromagnetic energy in a nonabsorbing medium are related by $\mathbf{S}' = w'\mathbf{v}'_{ph}$, Equation (1.19) can also be written in the form

$$w'_i v'_{ph,i} + w'_r v'_{ph,r} - w'_t v'_{ph,t} = 0. \tag{1.20}$$

We define the photon fluxes as

$$N'_i = \frac{S'_i}{\hbar\omega'_i}, \quad N'_r = \frac{S'_r}{\hbar\omega'_r}, \quad N'_t = \frac{S'_t}{\hbar\omega'_t}. \tag{1.21}$$

This definition has a clear meaning in the context of the quantization of the field, but in this context of classical electrodynamics it is clearly arbitrary.

As there is no change in frequency upon reflection and refraction, from Equation (1.19) one also trivially derives the conservation of the number of photons (Figure 1.2)

$$N'_i + N'_r - N'_t = 0. \tag{1.22}$$

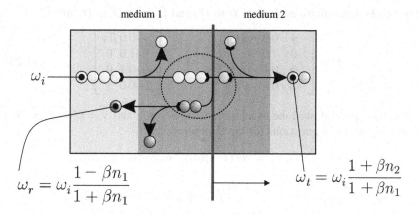

medium 1   medium 2

$$\omega_r = \omega_i \frac{1 - \beta n_1}{1 + \beta n_1} \qquad\qquad \omega_t = \omega_i \frac{1 + \beta n_2}{1 + \beta n_1}$$

FIGURE 1.2. Schematic representation of photon balance in the frame of reference at rest with media. Individual photons are pictorially represented as spheres. If we indicate with N_i', N_r', and N_t' the instantaneous fluxes of incoming, reflected, and refracted photons at the interface, we have, for the case in this figure, the ratio $\varrho'^2 = N_r'/N_i' = 2/3$ and $\tau'^2 \eta_1/\eta_2 = N_t'/N_i' = 1/3$, where ϱ'^2 and $\tau'^2 \eta_1/\eta_2$ are the classical power reflection and transmission coefficients. Indeed, the photon balance is equivalent to the classical power balance because of the uniformity of ω'.

The phase matching condition must be satisfied at the interface also in O because the phases are Lorentz invariant. Hence, there is only a single reflected and a single refracted wave also for the moving interface. To calculate the frequencies ω_i, ω_r, and ω_t and the k-vectors k_i, k_r, and k_t, as they are measured by the O observer, we have to transform the 4-vectors which contain the information about ω' and k' of the three waves from the O' to the O reference frame, according to the Lorentz transformation. The four-dimensional $\mathbf{k}'$-vectors are

$$\mathbf{K}_i' = \begin{bmatrix} \frac{\omega'}{c} \\ 0 \\ 0 \\ k_i' \end{bmatrix} \quad \mathbf{K}_r' = \begin{bmatrix} \frac{\omega'}{c} \\ 0 \\ 0 \\ k_r' \end{bmatrix} \quad \mathbf{K}_t' = \begin{bmatrix} \frac{\omega'}{c} \\ 0 \\ 0 \\ k_t' \end{bmatrix},$$

$$\mathbf{K}_i = \begin{bmatrix} \frac{\omega_i}{c} \\ 0 \\ 0 \\ k_i \end{bmatrix} \quad \mathbf{K}_r = \begin{bmatrix} \frac{\omega_r}{c} \\ 0 \\ 0 \\ k_r \end{bmatrix} \quad \mathbf{K}_t = \begin{bmatrix} \frac{\omega_t}{c} \\ 0 \\ 0 \\ k_t \end{bmatrix}.$$

(1.23)

Thus, we have to calculate

$$\mathbf{K}_i = \mathbf{L}' \cdot \mathbf{K}_i', \quad \mathbf{K}_r = \mathbf{L}' \cdot \mathbf{K}_r', \quad \mathbf{K}_t = \mathbf{L}' \cdot \mathbf{K}_t'. \tag{1.24}$$

The two Lorentz matrices, $\mathbf{L}$ from O to O', and $\mathbf{L}'$ from O' to O, are

$$
\mathbf{L} = \begin{bmatrix} \gamma & 0 & 0 & -\beta\gamma \\ 0 & 1 & 0 & 0 \\ 0 & 0 & 1 & 0 \\ -\beta\gamma & 0 & 0 & \gamma \end{bmatrix}, \quad
\mathbf{L}' = \begin{bmatrix} \gamma & 0 & 0 & \beta\gamma \\ 0 & 1 & 0 & 0 \\ 0 & 0 & 1 & 0 \\ \beta\gamma & 0 & 0 & \gamma \end{bmatrix}, \tag{1.25}
$$

where we have followed the usual definitions $\beta = v/c$ and $\gamma = 1/\sqrt{1-\beta^2}$. We finally obtain the expressions for the O observer as

$$
\omega_i = \gamma\omega'(1+\beta n_1), \ \omega_r = \gamma\omega'(1-\beta n_1), \ \omega_t = \gamma\omega'(1+\beta n_2),
$$

$$
k_i = \gamma\frac{\omega'}{c}(n_1+\beta), \ k_r = -\gamma\frac{\omega'}{c}(n_1-\beta), k_t = \gamma\frac{\omega'}{c}(n_2+\beta). \tag{1.26}
$$

The Doppler shift of the frequency of the reflected and refracted wave is then expressed as

$$
\omega_r = \omega_i \frac{1-\beta n_1}{1+\beta n_1}, \qquad \omega_t = \omega_i \frac{1+\beta n_2}{1+\beta n_1}. \tag{1.27}
$$

It is useful to find also the phase velocities of the incident, reflected and refracted waves for the O observer. From the general definition $v_{ph} = \omega/k$ it is

$$
v_{ph,i} = \frac{\omega_i}{k_i} = c\frac{1+\beta n_1}{n_1+\beta}, \quad v_{ph,r} = \frac{\omega_r}{k_r} = -c\frac{1-\beta n_1}{n_1-\beta}, \quad v_{ph,t} = \frac{\omega_t}{k_t} = c\frac{1+\beta n_2}{n_2+\beta}. \tag{1.28}
$$

It is immediate to verify that Equations (1.28) are a particular case of the dispersion shown as Equation (1.11). Equations (1.28) suggest a general definition for the observer O for the "effective" refractive index of a moving medium as $n_l = c/v_{ph}$. In general it is

$$
\vec{\overrightarrow{n_l}} = \frac{n_l+\beta}{1+n_l\beta}, \quad \overleftarrow{\overrightarrow{n_l}} = \frac{n_l-\beta}{1-n_l\beta}, \tag{1.29}
$$

where n_l is the refractive index of the medium at rest, and the left (right) equation applies to the case of phase velocity v_{ph} along (opposite to) the medium velocity v.

In order to gain physical insight into the expressions of Equations (1.28), it is customary to develop the Taylor expansions of $v_{ph}(\beta)$ to first order in β as

$$
v_{ph,i} \simeq = \frac{c}{n_1} + v\left(1-\frac{1}{n_1^2}\right),
$$

$$
v_{ph,r} \simeq = \frac{c}{n_1} - v\left(1-\frac{1}{n_1^2}\right), \tag{1.30}
$$

$$
v_{ph,t} \simeq = \frac{c}{n_2} + v\left(1-\frac{1}{n_2^2}\right),
$$

where, besides the c/n_l term — the "still" medium term appearing also in Equations (1.3) — an additional term appears, which is linear in v, referred to as *Fresnel drag*. Because of the Fresnel drag, the motion of the medium *pulls* the light along the direction of the medium velocity v. For example, for positive velocity v in Figure 1.1, the phase velocity of incident and refracted light increases, whereas it decreases for the reflected light. In particular, the Fresnel drag becomes critical when $|\beta| > n_l^{-1}$ i.e., when the velocity of the medium exceeds the velocity of light in the medium at rest. The sign of frequency and velocity change. While the latter effect is intuitively understandable, the change of sign of frequency has a deeper meaning, which is physically interpreted in the sketch shown in Figure 1.3.

We now consider the transformation of the fields. At normal incidence, polarizations of wave are degenerate. If we assume that in O' $\mathbf{E}'$ is directed as $\mathbf{u}'_x$, then the second of Equations (1.1) implies that $\mathbf{B}'$ is directed as $\mathbf{u}'_y$, and Equations (1.3) imply that $\mathbf{D}'$ and $\mathbf{H}'$ are also respectively directed as $\mathbf{u}'_x$ and $\mathbf{u}'_y$. The Lorentz transformations (1.2) in the particular case of translational velocity along z — Equations (1.25) — maintain the same directions, respectively, also for $\mathbf{E}$, $\mathbf{B}$, $\mathbf{D}$,

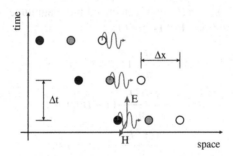

FIGURE 1.3. In this schematic plot, we represent a one-dimensional medium, along the axis of space, as a periodic lattice of circles (only three circles are shown for simplicity). The circles are colored differently to help track their position in time. The effect of medium on light propagation, which is described by its refractive index n_l, is represented in a simplified way as a process of successive absorptions and reemissions of light photons. We assume that the process reproduces itself periodically in time. In a period Δt, an emission, a travel to the adjacent (moving) lattice site, and an absorption happen. For $|\beta| > n_l^{-1}$, the velocity of moving medium is larger than the phase velocity of light. In the figure, we consider this condition with the light counterpropagating with respect to the medium. The figure shows that the sign of Poynting vector $\mathbf{S}$ does not agree with the sign of the group velocity of the perturbation (which in the present nondispersive medium coincides with the phase velocity). Since $\mathbf{S} = w \mathbf{v}_{ph}$ holds, where w is the energy density (1.33) and $\mathbf{v}_{ph}$ is the velocity of the wave (1.28), this also explains the negative sign of the energy density w.

and **H**. The transformed field components in O are

$$E_i = \gamma E_i'(1+\beta n_1), \quad E_r = \gamma \varrho' E_i'(1-\beta n_1), \quad E_t = \gamma \tau' E_i'(1+\beta n_2),$$

$$B_i = \frac{\gamma E_i'}{c}(\beta + n_1), \quad B_r = \frac{\gamma \varrho' E_i'}{c}(\beta - n_1), \quad B_t = \frac{\gamma \tau' E_i'}{c}(\beta + n_2),$$

$$D_i = \frac{\gamma E_i'}{c\eta_1}(n_1 + \beta), \quad D_r = \frac{\gamma \varrho' E_i'}{c\eta_1}(n_1 - \beta), \quad D_t = \frac{\gamma \tau' E_i'}{c\eta_2}(n_2 + \beta),$$

$$H_i = \frac{\gamma E_i'}{\eta_1}(n_1\beta + 1), \quad H_r = \frac{\gamma \varrho' E_i'}{\eta_1}(n_1\beta - 1), \quad H_t = \frac{\gamma \tau' E_i'}{\eta_2}(n_2\beta + 1),$$

$$(1.31)$$

where E_i' is the electric field in the reference frame O'.

From Equations (1.31), two general results should be pointed out. First, the ratio $E/H = \eta$ for any plane wave does not depend on the state of motion of the dielectric. This is not trivially expected from the Lorenz transformation: for example, the ratio E/B does depend on the state of motion. Second, since $-1 < \beta < 1$, the two fields B and D never change sign, but the fields E and H do: this happens for waves traveling in the opposite direction to the translation of the medium, in the critical condition ($|\beta| > n_i^{-1}$).

In Equations (1.31) we have only calculated the magnitude of the fields. The Poynting vector $\mathbf{S} = \mathbf{E} \times \mathbf{H}$ in O is

$$\mathbf{S}_i = +\frac{\gamma^2 E_i'^2}{\eta_1}(1+\beta n_1)^2 \mathbf{u}_z,$$

$$\mathbf{S}_r = -\frac{\gamma^2 \varrho'^2 E_i'^2}{\eta_1}(1-\beta n_1)^2 \mathbf{u}_z, \qquad (1.32)$$

$$\mathbf{S}_t = +\frac{\gamma^2 \tau'^2 E_i'^2}{\eta_2}(1+\beta n_2)^2 \mathbf{u}_z.$$

The sign of the Poynting vector is invariant for any value of β because the change of sign in the critical condition $|\beta| > n_i^{-1}$ occurs simultaneously for **E** and **H**.

The electromagnetic energy density $w = 1/2(\mathbf{E} \cdot \mathbf{D} + \mathbf{B} \cdot \mathbf{H})$ in O is

$$w_i = \frac{\gamma^2 E_i'^2}{c\eta_1}(1+\beta n_1)(n_1 + \beta),$$

$$w_r = \frac{\gamma^2 \varrho'^2 E_i'^2}{c\eta_1}(1-\beta n_1)(n_1 - \beta), \qquad (1.33)$$

$$w_t = \frac{\gamma^2 \tau'^2 E_i'^2}{c\eta_2}(1+\beta n_2)(n_2 + \beta),$$

where we notice that in critical condition $|\beta| > n_l^{-1}$ the energy density becomes negative, but the relationship $\mathbf{S} = w\mathbf{v}_{ph}$ still holds, as for media at rest. We provide an interpretation of the result of negative energy in Figure 1.3.

In the laboratory frame, to prove the conservation of energy, one must take into account the mechanical energy which is transferred to the material system by the electromagnetic pressure applied by the waves to the interface. The momentum density (momentum per unit volume) $\mathbf{g}$ carried by the plane wave is [11]

$$\mathbf{g} = \mathbf{D} \times \mathbf{B} = \frac{\mathbf{S}}{v_{ph}^2}, \tag{1.34}$$

where the last equality is a consequence of the relationships $\mathbf{D} = \mathbf{E}/v_{ph}$ and $\mathbf{B} = \mathbf{H}/v_{ph}$, which hold for any state of inertial motion, as directly derived from Equations (1.31). The mechanical power performed by the electromagnetic wave on the interface per unit area is

$$W = \mathbf{g} \cdot \mathbf{v} \tag{1.35}$$

diminished by the momentum per unit time per unit area stored in the space created by the interface motion ($\mathbf{g}v$). Since $\mathbf{g}$ and $\mathbf{S}$ are related by Equation (1.34), one can show that the mechanical energy flux is equal to the excess energy per unit time per unit area ΔE injected to the interface by the electromagnetic waves, which is the balance of Poynting vectors of the waves, diminished by the energy per unit time per unit area stored in the space created by the interface motion:

$$\begin{aligned}
\Delta E' &= w_i(v_{ph,i} - \beta c) + w_r(v_{ph,r} - \beta c) - w_t(v_{ph,t} - \beta c) \\
&= \frac{E_i'^2}{\eta_1}(1 + \beta n_1) - \frac{\varrho'^2 E_i'^2}{\eta_1}(1 - \beta n_1) - \frac{\tau'^2 E_i'^2}{\eta_2}(1 + \beta n_2) \neq 0.
\end{aligned} \tag{1.36}$$

The last line of Equation (1.36), when compared to Equations (1.33) and (1.3), suggests to repeat the calculation for the photon flux instead of the energy flux. We must divide the terms of Equation (1.36) by $\hbar\omega$, where ω is the respective frequency of each term, as calculated by Equations (1.3). Thus, we calculate the photon balance as the difference ΔN between the incoming and outgoing photon fluxes of the waves, diminished by the stored photons in the space created by the interface motion:

$$\begin{aligned}
\Delta N &= \frac{w_i}{\hbar\omega_i}(v_{ph,i} - \beta c) + \frac{w_r}{\hbar\omega_r}(v_{ph,r} - \beta c) - \frac{w_t}{\hbar\omega_t}(v_{ph,t} - \beta c) \\
&= \frac{E_i'^2}{\gamma\hbar\omega'\eta_1}\frac{(1 + \beta n_1)}{(1 + \beta n_1)} - \frac{\varrho'^2 E_i'^2}{\gamma\hbar\omega'\eta_1}\frac{(1 - \beta n_1)}{(1 - \beta n_1)} - \frac{\tau'^2 E_i'^2}{\gamma\hbar\omega'\eta_2}\frac{(1 + \beta n_2)}{(1 + \beta n_2)} \\
&= \frac{E_i'^2}{\gamma\hbar\omega'}\left(\frac{1}{\eta_1} - \frac{\varrho'^2}{\eta_1} - \frac{\tau'^2}{\eta_2}\right) = 0,
\end{aligned} \tag{1.37}$$

where we have made use of Equation (1.19). The interpretation of this last interesting result is that the interface between the two dielectrics acts as a

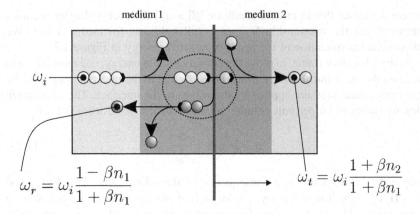

medium 1   medium 2

$$\omega_i$$

$$\omega_r = \omega_i \frac{1 - \beta n_1}{1 + \beta n_1}$$

$$\omega_t = \omega_i \frac{1 + \beta n_2}{1 + \beta n_1}$$

FIGURE 1.4. Schematic representation of photon balance in a frame of reference O in motion with respect to the frame O' of media (axis conventions are given in Figure 1.1). The interface moves with velocity $v\mathbf{u}_z$ in O as indicated by the arrow. Individual photons are pictorially represented as spheres. The ratios of instantaneous fluxes of photons — reflected versus incoming and refracted versus incoming — *at the interface* (i.e., inside the dashed circle) are *invariant* with respect to the status of motion (assumed inertial), and are equal to the power reflection and transmission coefficients ϱ'^2 and $\tau'^2 \eta_1/\eta_2$ as calculated in the frame at rest with the interface. In this example, $\varrho'^2 = 2/3$ and $\tau'^2 \eta_1/\eta_2 = 1/3$. The interpretation of this interesting result is that the interface between the two dielectrics acts as a number-conserving and invariant photon converter, regardless of its state of (inertial) motion. This joint invariance and conservation law is fully contained inside classical electrodynamics.

number-conserving and invariant photon converter, regardless of its state of (inertial) motion, in quantitative consistency with both Doppler shifts and the conservation of energy. This joint invariance and conservation law is fully contained inside classical electrodynamics. The interpretation is schematically shown in Figure 1.4.

1.4 Traveling Interface Case

The physical configuration discussed in this section is schematically represented in Figure 1.5: the media are at rest *in the frame* O and the electromagnetically created interface is *moving* at a velocity v in the z-direction. On the other hand, *the interface* is at rest in the frame O'. Since the media are now in motion with z-velocity $-v$ with respect to O', the calculation of frequency and k-vectors can be carried on as in the previous paragraphs, provided that we use the effective refractive index — as defined in Equations (1.29) — instead of the index of the medium at rest.

The frequency is still conserved because the phase matching condition still applies (it depends on the motion condition of the interface only, which is at

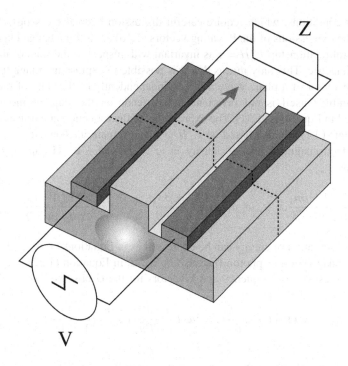

FIGURE 1.5. Schematic representation of a *traveling wave refractor*. Light travels in a waveguide, whose core exhibits some electro-optic effect. The electro-optic effect is controlled by two side electrodes, which are terminated at the end side by a suitable impedance Z. For example, if Z matches the characteristic impedance of the electrode line, the traveling signal exhibits no reflection at Z. The driving signal is provided by a microwave generator V connected between the other sides of the electrodes, which inject a driving signal for the electro-optic effect. The wavefront of the microwave (dashed line) represents a moving interface between two portions of the medium, each with different refractive index. Hence, it can be viewed as an interface in relative motion with respect to two effective media. In order to focus on the effect of motion only, we make the simplifying assumption that the refractive index is everywhere constant in the core, except at the microwave interface, where it exhibits a discontinuity.

rest). Thus, it is again $\omega'_i = \omega'_r = \omega'_t = \omega'$ as in Section (1.3). All the equations of Section (1.3) that depend on the index of refraction must now be modified. Equations (1.3) must be replaced by

$$k'_i = \frac{\omega'}{c} \frac{n_1 - \beta}{1 - \beta n_1}, \quad k'_r = -\frac{\omega'}{c} \frac{n_1 + \beta}{1 + \beta n_1}, \quad k_t = \frac{\omega'}{c} \frac{n_2 - \beta}{1 - \beta n_2},$$

$$v'_{ph,i} = c \frac{1 - \beta n_1}{n_1 - \beta}, \quad v'_{ph,r} = -c \frac{1 + \beta n_1}{n_1 + \beta}, \quad v'_{ph,t} = c \frac{1 - \beta n_2}{n_2 - \beta}.$$

(1.38)

We assume that $\beta \leq n_1^{-1}$ and $\beta \leq n_2^{-1}$. This assumption means that the interface is traveling at slower velocity than any phase velocity. We do not consider here different conditions, which require careful discussion beyond the scope of this work. The expressions of the Poynting vectors (1.16) are left unchanged because the general relationship $E/H = \eta$ is invariant with respect to the state of motion of the dielectric. This only means that it is possible to express the magnitude of Poynting vector of a plane wave as E^2/η independently of the state of motion: the magnitude itself is not invariant and depends on the state of motion as calculated in Equations (1.31). The expressions of the reflection and transmission coefficients (3.1) are also based on $E/H = \eta$ and remain unchanged.

The electromagnetic energy density $w' = 1/2(\mathbf{E}' \cdot \mathbf{D}' + \mathbf{B}' \cdot \mathbf{H}')$ in O' is, for each wave,

$$w_i' = \frac{E_i'^2}{c\eta_1}\frac{1-\beta n_1}{n_1-\beta}, \qquad w_r' = \frac{\varrho'^2 E_i'^2}{c\eta_1}\frac{1+\beta n_1}{n_1+\beta}, \qquad w_t' = \frac{\tau'^2 E_i'^2}{c\eta_2}\frac{1-\beta n_2}{n_2-\beta}. \quad (1.39)$$

The conservation of energy can be expressed as in Equations (1.19) and (1.20) and the conservation of photon flux also holds, as in Equation (1.22).

The expressions of frequency and k-vectors for the O observer are

$$\omega_i = \gamma\omega'(1+\beta\frac{n_1-\beta}{1-\beta n_1}) = \gamma\omega'(\frac{1-\beta^2}{1-\beta n_1}) = \frac{\omega'}{\gamma}(\frac{1}{1-\beta n_1}),$$

$$\omega_r = \frac{\omega'}{\gamma}(\frac{1}{1+\beta n_1}), \qquad \omega_t = \frac{\omega'}{\gamma}(\frac{1}{1-\beta n_2}),$$

$$k_i = \gamma\frac{\omega'}{c}(\frac{n_1-\beta}{1-\beta n_1}+\beta) = \gamma\frac{\omega'n_1}{c}(\frac{1-\beta^2}{1-\beta n_1}) = \frac{\omega'n_1}{\gamma c}(\frac{1}{1-\beta n_1}),$$

$$k_r = -\frac{\omega'n_1}{\gamma c}(\frac{1}{1+\beta n_1}), \qquad k_t = \frac{\omega'n_2}{\gamma c}(\frac{1}{1-\beta n_2}). \quad (1.40)$$

If we now calculate the phase velocity ($v_{ph} = \omega/k$), we find

$$v_{ph,i} = \frac{\omega_i}{k_i} = \frac{c}{n_1}, \qquad v_{ph,r} = \frac{\omega_r}{k_r} = -\frac{c}{n_1}, \qquad v_{ph,t} = \frac{\omega_t}{k_t} = \frac{c}{n_2}, \quad (1.41)$$

consistent with the fact that the media are still for the observer O.

The Doppler shift of the frequency of the reflected and refracted wave is then expressed as

$$\omega_r = \omega_i\frac{1-\beta n_1}{1+\beta n_1}, \qquad \omega_t = \omega_i\frac{1-\beta n_1}{1-\beta n_2}, \quad (1.42)$$

where it is interesting to note that the Doppler shift of the reflected wave is unchanged with respect to Equation (1.27), but the shift of the transmitted wave is different.

The transformed field components in O are

$$E_i = \frac{E_i'}{\gamma} \frac{1}{(1-\beta n_1)}, \qquad E_r = \frac{\varrho' E_i'}{\gamma} \frac{1}{(1+\beta n_1)}, \qquad E_t = \frac{\tau' E_i'}{\gamma} \frac{1}{(1-\beta n_2)},$$

$$B_i = \frac{E_i'}{\gamma c} \frac{n_1}{(1-\beta n_1)}, \qquad B_r = -\frac{\varrho' E_i'}{\gamma c} \frac{n_1}{(1+\beta n_1)}, \qquad B_t = \frac{\tau' E_i'}{\gamma c} \frac{n_2}{(1-\beta n_2)},$$

$$D_i = \frac{E_i'}{\gamma \eta_1 c} \frac{n_1}{(1-\beta n_1)}, \qquad D_r = \frac{\varrho' E_i'}{\gamma \eta_1 c} \frac{n_1}{(1+\beta n_1)}, \qquad D_t = \frac{\tau' E_i'}{\gamma \eta_2 c} \frac{n_2}{(1-\beta n_2)},$$

$$H_i = \frac{E_i'}{\gamma \eta_1} \frac{1}{(1-\beta n_1)}, \qquad H_r = -\frac{\varrho' E_i'}{\gamma \eta_1} \frac{1}{(1+\beta n_1)}, \qquad H_t = \frac{\tau' E_i'}{\gamma \eta_2} \frac{1}{(1-\beta n_2)},$$

$$\tag{1.43}$$

where the symbols ϱ' and τ' are, as already stated, unchanged from Equations (3.1).

The Poynting vectors $\mathbf{S} = \mathbf{E} \times \mathbf{H}$ in O are

$$\mathbf{S}_i = +\frac{E_i'^2}{\gamma^2 \eta_1}(1-\beta n_1)^{-2} \mathbf{u}_z,$$

$$\mathbf{S}_r = -\frac{\varrho'^2 E_i'^2}{\gamma^2 \eta_1}(1+\beta n_1)^{-2} \mathbf{u}_z, \tag{1.44}$$

$$\mathbf{S}_t = +\frac{\tau'^2 E_i'^2}{\gamma^2 \eta_2}(1-\beta n_2)^{-2} \mathbf{u}_z.$$

The electromagnetic energy density $w = 1/2(\mathbf{E} \cdot \mathbf{D} + \mathbf{B} \cdot \mathbf{H})$ in O is

$$w_i = \frac{E_i'^2 n_1}{c\gamma^2 \eta_1}(1-\beta n_1)^{-2},$$

$$w_r = \frac{\varrho'^2 E_i'^2 n_1}{c\gamma^2 \eta_1}(1+\beta n_1)^{-2}, \tag{1.45}$$

$$w_t = \frac{\tau'^2 E_i'^2 n_2}{c\gamma^2 \eta_2}(1-\beta n_2)^{-2}.$$

Even though we have limited the analysis to $\beta \leq n_1^{-1}$ and $\beta \leq n_2^{-1}$, we mention that the energy density as resulting from Equations (1.45) does not become negative under any condition, consistent with the fact that media are at rest in the observer frame.

Following the same reasoning as in Section 1.3, we now obtain, in place of Equation (1.37),

$$
\begin{aligned}
\Delta N &= \frac{w_i}{\hbar\omega_i}(v_{ph,i}-\beta c) + \frac{w_r}{\hbar\omega_r}(v_{ph,r}-\beta c) - \frac{w_t}{\hbar\omega_t}(v_{ph,t}-\beta c) \\
&= \frac{E_i'^2}{\gamma\hbar\omega'\,\eta_1}\frac{(1-\beta n_1)^2}{(1-\beta n_1)^2} - \frac{\varrho'^2 E_i'^2}{\gamma\hbar\omega'\,\eta_1}\frac{(1+\beta n_1)^2}{(1+\beta n_1)^2} - \frac{\tau'^2 E_i'^2}{\gamma\hbar\omega'\,\eta_2}\frac{(1-\beta n_2)^2}{(1-\beta n_2)^2} \\
&= \frac{E_i'^2}{\gamma\hbar\omega'}\left(\frac{1}{\eta_1} - \frac{\varrho'^2}{\eta_1} - \frac{\tau'^2}{\eta_2}\right) = 0.
\end{aligned}
\tag{1.46}
$$

Because of Equation (1.19), the joint invariance and conservation law $\Delta N = 0$ illustrated in Figure 1.4 is valid also for this case. Also in this traveling interface case, there is a sizable Doppler shift (1.42) of both the reflected and the transmitted waves.

1.5 Conclusions

In this chapter we have discussed some interesting consequences of Maxwell's equations in moving media: the moving interface between two inertial dielectrics acts as a number-conserving and invariant photon converter. Although the calculations have been performed using classical electrodynamics, the results have a clear interpretation in terms of conservation of number of energy quanta if the photon flux is calculated by dividing the classical power fluxes of incident, reflected, and refracted waves by the quantities $\hbar\omega_{i,r,t}$, where $\hbar$ is the reduced Planck constant and $\omega_{i,r,t}$ are the corresponding, Doppler shifted, frequencies.

Two cases have been considered: the standard, *material interface* one, where the two media are rigidly moving in the laboratory frame, and a novel, *traveling interface* one, where the medium is at rest, and the interface is created by a traveling electromagnetic pulse via the electro-optic effect. The Doppler shifts of the reflected and the transmitted waves have been calculated for the two cases: while the shifts of the reflected wave are the same in the two cases, those of the transmitted wave differ. As it does not require mechanical objects moving at relativistic speed, the traveling interface system appears as most suited to produce large wavelength shifts.

From the point of view of basic physics, it appears as very promising in view of studies of interfaces at ultrarelativistic speeds, where the interface velocity is larger than the phase and group velocities in the media. This regime is very interesting as it opens the way toward experimental studies of optical black holes, and of the related quantum effects.

Acknowledgments. The help of L. Pavesi, F. Riboli, A. Recati, I. Carusotto, S. Prezioso, and M. Ghulinyan is acknowledged.

References

1. L. Pavesi and D. J. Lockwood, *Silicon Photonics*, Springer, Berlin (2004).
2. T. Durhuus, B. Mikkelsen, C. Joergensen, S. L. Danielsen, and K. E. Stubkjaer, *J. Lightwave Technol.* **14**, 942–953 (1996).
3. S. J. B. Yoo, *J. Lightwave Technol.* **14**, 955–966 (1996).
4. M. Lampe, E. Ott, and J. H. Walker, *Phys. Fluids* **21**, 42–54 (1989).
5. I. Scherbatko, *Opt. Quantum Electron.* **31**, 965–979 (1999).
6. A. Sommerfeld, *Electrodynamics*, Academic Press, New York (1952).
7. J. D. Jackson, *Classical Electrodynamics*, Wiley, New York (1962).
8. H. Minkowski, *Göttinger Nachr.*, p. 53 (1908).
9. S. J. Orfanidis, *Electromagnetic Waves and Antennas*. www.ece.rutgers.edu/ orfanidi/ ewa, 2004.
10. J. M. Saca, *J. Mod. Opt.* **36**, 1367–1376 (1989).
11. W. Greiner, *Classical Electrodynamics*, Springer, Berlin (1998).

2

Evanescent Waves as Nanoprobes for Surfaces and Interfaces: From Waveguide Technology to Sensor Application

Silvia Mittler[1] and Bernhard Menges[2]

[1] Department of Physics and Astronomy, The University of Western Ontario,
London, Ontario N6A 3K7, Canada
smittler@uwo.ca
[2] Max-Planck-Institut für Polymerforschung, Ackermann Weg
10, 55128 Mainz, Germany

2.1 Introduction

This chapter does not deal with a system which is nanoscopic in all three dimensions of space. A classical nano-object with nanoscopic dimensions in all three dimensions would be, for instance, a nanoparticle which can exhibit quantum effects due to its size. In the case of an evanescent wave at a surface of an optical waveguide we deal with a nanoscopic effect in one dimension only: perpendicular to the propagation direction of the electromagnetic wave. Both other dimensions are macroscopic and in the theory typically treated as infinite. But because of this seminanoscopic circumstance at an "optical surface" we speak about Surface Nanophotonics. The future will definitely show how these one-dimensional nanodevices can be combined with two- or three-dimensional nanostructures to foster new physical effects and new applications.

2.2 Waveguides

An optical waveguide is in principle a very simple structure. In its simplest form it consists of one sheet of transparent dielectric material with a thickness in the range of a micrometer. Because it is surrounded by air having a refractive index $n(\text{air}) = 1$, light propagating in the thin layer of the dielectric material with a refractive index larger than 1 [n(waveguide dielectric material) > 1] can experience total internal reflection at the dielectric–air interface when it is incident above the critical angle of total internal reflection $\theta_{critical}$ [$\sin\theta_{critical} = n(\text{air})/n(\text{waveguide dielectric material})$] [1–3].

When the light faces this total internal reflection on both sides of the dielectric sheet, the light is trapped and can only propagate by zigzagging between both interfaces within and along the sheet (Figure 2.1a). It is then called a guided wave or beam and the dielectric sheet an optical waveguide. Not all angles above the critical angle of total internal reflection will lead to an efficient guiding of the light. Only certain angles of reflection can lead to an efficient propagation of the light along the guide which leads to constructive interference. Therefore, a mode structure exists with specific angles of reflection.

If we have a closer look at the dielectric–air interface when light is reflected there totally, we find that with every reflection the beam experiences an offset (a little shift in position toward the main propagation direction), called the Goos-Hänchen shift [1, 4–7]. This shift occurs because the total reflection is not happening completely at the dielectric–air interface, but partly in the air. Outside of the dielectric sheet a "film of light" decaying exponentially away in intensity from the interface is found. This is called the evanescent field.

If one treats this three-layer problem (air–dielectric transparent sheet–air) with electromagnetic theory, Maxwell's equations lead in combination with the boundary conditions at both interfaces to two wave equations, one for each polarization direction (s- and p-polarized) and their solutions describe standing

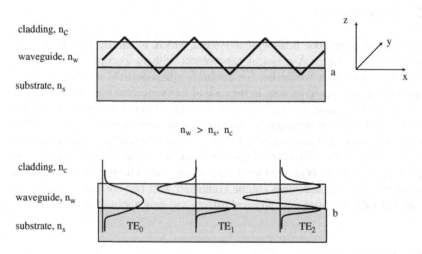

FIGURE 2.1. (a) Asymmetric three layer waveguide architecture with the total internally reflected beam zigzagging along the waveguide. The Goos-Hänchen shift is denoted by the intrusion of the beam into the surrounding materials. (b) Here the field distributions of the first three s-polarized modes are shown. The numbering of the modes is in accordance with their appearing by increasing the waveguide thickness. The index number also signifies the number of zero-field crossings in the field distribution.

waves across the dielectric sheet [4–9]. Depending on the refractive index of the dielectric material and the thickness of the dielectric sheet, various standing wave solutions can be found, all of them showing an oscillating field distribution within the dielectric sheet and evanescent fields on both sides outside the structure (Figure 2.1b). These various solutions are the modes, mentioned above as belonging to specific angles of reflection. In the symmetric case the field distribution is symmetric.

Each of these modes has a characteristic wave vector β_m which is aligned parallel to the waveguide sheet and points into the propagation direction. The effective refractive index N_m for each mode is related to the wave vector by: $\beta_m = 2\pi N_m/\lambda$, with λ the wavelength of the propagating light. In practice it is difficult to handle sheets of micrometer thickness, so-called free-standing waveguides. Therefore, the dielectric waveguide material is usually located on top of a substrate. In order to keep the condition of total internal reflection in the guide, the substrate needs to have a lower refractive index than the waveguide. Such a structure is called a slab waveguide. The field distribution here is asymmetric because of the asymmetric refractive index distribution. Figure 2.1 shows such an asymmetric waveguide structure with a possible zigzag path including the Goos-Hänchen shift in panel a and the theoretically calculated field distribution for three modes in panel b.

Alternative geometries are the fiber optical guides with a cylindrical geometry, trapping the light into a cylinder, basically a channel, where total internal reflection happens in "two" dimensions, confining the beam laterally into the cylinder (Figure 2.2). This kind of stripe or channel waveguides can also be defined in slab waveguides if a lateral refractive index distribution is present,

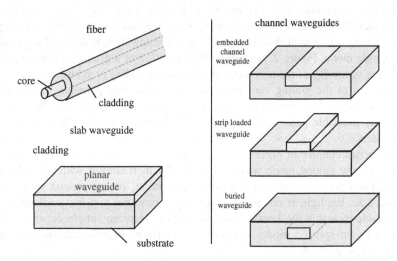

FIGURE 2.2. (Left) The two fundamental waveguide geometries: the fiber and the planar slab waveguide. (Right) Three types of channel waveguides: embedded channel waveguide, strip loaded waveguide, and buried waveguide.

allowing total internal reflection of the light back into the stripe from two sides, in addition to the top–bottom confinement of the dielectric sheet itself: a high refractive index material in stripe form embedded in a low refractive index material. Figure 2.2 shows some typical waveguide geometries: the fiber, the slab waveguide, and three classes of channel waveguides. These channel waveguides now allow for various integrated optical devices; light can be distributed from one channel into two or more channels (a device called a splitter), or combined from two channels into one.

To summarize: a waveguide is a thin film with a "high" refractive index located on a "low" refractive index substrate guiding light in the form of modes with specific field distributions in two possible polarization directions (s-polarization leads to TE modes and p-polarization to TM modes), always showing an evanescent field outside the structure.

Many applications of waveguides are in telecommunications, where the guides are used to bring information in the form of modulated intensities or light pulses to specific targets. Long-distance communication is usually achieved with the use of optical fibers, whereas signal distribution is performed by integrated optical channel waveguides. In all of these applications the guiding effect itself is used [10–12].

In the field of integrated optical sensors, it is not the guiding effect itself that is used, but interactions in the evanescent field present on the surface of the device, or changes in the optical constants within the waveguide material leading to changes in the field distribution.

In order to understand how these changes—in the evanescent field or the waveguide itself—can be identified, we discuss how the light can be launched into such thin film structures. Here, basically two classical techniques are available: (1) methods which couple the light by matching the wave vector $\mathbf{k}$ of the incident light with the wave vector $\boldsymbol{\beta}_m$ of a waveguide mode, and (2) coupling by matching the spatial distribution of the incoming light beam to that of the waveguide mode. Figure 2.3a explains the operation of a grating coupler where the wave vector $\boldsymbol{\beta}_m$ of the mth mode is matched by adding or subtracting integer multiples n of the grating vector $\mathbf{G}$; the latter is oriented perpendicular to the grating lines and has a length $G = 2\pi/\Lambda$, with Λ the grating spacing. The experimentally available parameter to adjust for resonant coupling is then the angle of incidence θ.

The prism coupler in Figure 2.3b matches the wave vector of the incoming light with the guided mode wave vector $\boldsymbol{\beta}_m$ by increasing it with the help of the higher refractive index of the prism material n_p in comparison to air. In Figure 2.3c, the light is coupled by focusing it onto the polished endface of a waveguide to match the lateral field distribution. Here no simple experimental selectivity for specific modes is possible, because no experimental adjustment parameter is available.

Figure 2.4 shows the scheme of the experimental setup for waveguide characterization. A laser beam at a given wavelength is incident on a coupling grating. The transmitted intensity behind the waveguide is monitored while the angle of

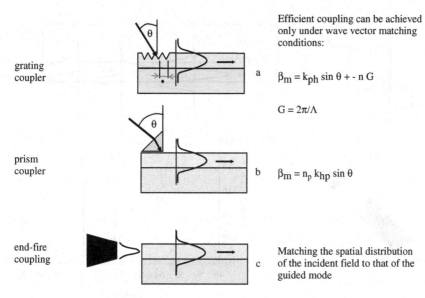

grating
coupler

Efficient coupling can be achieved
only under wave vector matching
conditions:

a

$$\beta_m = k_{ph} \sin \theta + - n\,G$$

$$G = 2\pi/\Lambda$$

prism
coupler

b

$$\beta_m = n_p\, k_{hp} \sin \theta$$

end-fire
coupling

c

Matching the spatial distribution
of the incident field to that of the
guided mode

FIGURE 2.3. The three classical coupling techniques: (a) grating coupler, (b) prism coupler, (c) end-fire coupling.

incidence is scanned. Each successfully coupled mode is found as an intensity dip in the intensity – angle of incidence scan, because when the light is coupled to the waveguide, it is missing in transmission. The highest resonance angle found is always the 0th mode. In the left-hand corner the thickness and refractive index information is given for that particular mode spectrum at $\lambda = 632.8\,$nm. The resonance angles θ_m are taken from such a mode spectrum and converted into the effective refractive indices N_m for the various modes:

$$\text{In the case of grating coupling}: N_m = n(\text{air}) \sin \theta_m \pm n \frac{\lambda}{\Lambda}$$

$$\text{In the case of prism coupling}: N_m = \sin(\theta_m) \cos \varepsilon + \sqrt{n_p{}^2 - \sin^2 \theta_m}\ \sin \varepsilon$$

with ε the prism angle (see Figure 2.3). As long as at least two modes are found and the optical data of the substrate are known, the thickness and the refractive index of the waveguide material can be obtained with the help of the effective refractive indices of the modes and an iterative computer routine based on a procedure proposed by Ulrich and Torge in 1973 [13–15].

Because two polarization directions can be used independently to characterize the waveguide, anisotropic refractive indices, arising from anisotropic waveguide fabrication processes, can easily be detected. The simplest form of sensor then involves monitoring changes in the thickness or the refractive indices in the waveguide itself due to any physical or chemical treatment, such as photo-bleaching, aging, milling, wear, or swelling, (see example below).

refractive index and thickness profile

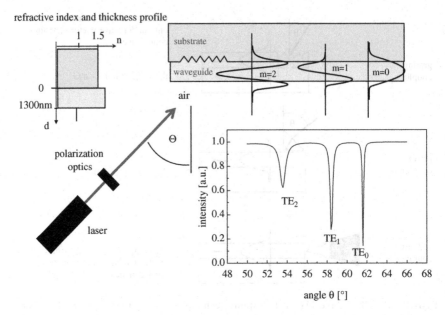

FIGURE 2.4. Measurement scheme for waveguide characterization with respect to waveguide refractive index and waveguide thickness in s- and p-polarization. The angle of incidence is scanned and the transmitted intensity of a laser incident on the grating is monitored. If a resonance angle is approached the laser beam is coupling into the waveguide structure, so that the light is missing in transmission. Therefore, resonance dips are found in the intensity–angle of incidence scan. The inset shows the waveguide architecture with respect to thicknesses and refractive indices.

The materials used for these kinds of structures vary from classical glasses, engineered glasses, or semiconductors all the way to organic materials, such as polymers, and combinations thereof. Fabrication technologies are also spread along a huge variety of deposition or material modification methods. Two strategies are followed for waveguide fabrication: the deposition of the waveguide material on top of a transparent substrate or the enhancement of the refractive index in a substrate. The deposition techniques include simple approaches like spin coating (polymers and solgel systems) [16–18] time-consuming methods like Langmuir-Blodgett (LB) film transfer (small organic molecules or polymers leading to intrinsic order) [19–22], extruding (insoluble polymers) [23, 24], and highly sophisticated CVD processes (glasses and semiconductors) [25–28]. The material alteration methods rely on the implantation or exchange of ions with a high polarizability to enhance the refractive index in the material [29]. Typical methods are ion exchange [30, 31] and ion implantation [32].

In all of these technologies one aim is to fabricate waveguides with low intrinsic losses. The waveguide losses are due to electronic absorptions of the material and scattering losses. Therefore, not all materials are suitable for all optical wavelengths. The regime of transparency dictates the possible operational wavelengths. Scattering can occur on the surface due to roughness, and internally in the waveguide due to refractive index fluctuations such as grain boundaries or orientation fluctuations. Therefore, the waveguide material should be perfectly glassy or alternatively a single crystal [33–35].

The evanescent field is used in integrated optics sensor technology as described by Figure 2.5. A well characterized waveguide is used as a substrate for an adlayer. In the first approximation, let us consider the adlayer as the material to be detected and the waveguide completely characterized. The asymmetric three-layer system becomes a four-layer system, still with two unknowns: the thickness and the refractive index of the adlayer. This more complicated structure as well as five-layer and multilayer structures are theoretically treated in various books [4, 14]. Typically a matrix formalism based on the Fresnel equations [4, 14, 36] is used to calculate the field contributions reflected and refracted at each interface.

As seen in Figure 2.5, the adsorption of the adlayer leads to a shift of a waveguide mode to a higher resonance angle, leading to an enhanced effective refractive index N of that mode. The sensitivity S of such a waveguide sensor can be defined as the change in effective refractive index of the mode δN with thickness increment δd of the adlayer at a fixed refractive index: $S = \frac{\delta N}{\delta d}$. It turns out theoretically and experimentally that this sensitivity S is a function

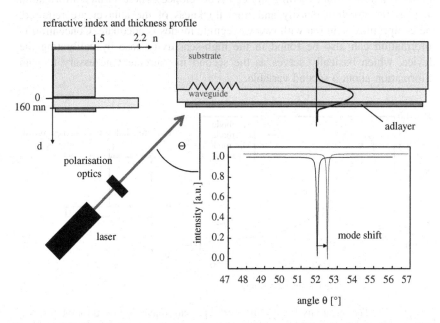

FIGURE 2.5. A well-characterized waveguide carries an additional adlayer. Due to the adlayer the waveguide mode resonant coupling angle is shifted to a higher angle of incidence. The inset shows the waveguide architecture with respect to thicknesses and refractive indices.

of the refractive index contrast $\Delta n = |n(\text{waveguide})-n(\text{medium on top})|$ of the waveguide and the medium on top, as well as on the thickness of the actual waveguide [37, 38]. Figure 2.6 shows the calculated sensitivity S for an asymmetric Ta_2O_5 waveguide having a refractive index of 2.2, with an adlayer at $n = 1.5$ at 632.8 nm, with increasing waveguide thickness. First, one sees that a minimum waveguide thickness is required for the TE and TM modes to exist in this asymmetric structure. This minimum thickness is called the cutoff thickness. The cutoff thickness is smaller for the TE mode than the TM mode at a given mode number and increases with increasing mode number. The sensitivity maxima are always found very close to the cutoff thicknesses and the sensitivity peak heights decrease with increasing mode number or waveguide thickness, respectively.

The sensor designer has to make a choice on how to design a sensor: (1) use a waveguide with maximal possible sensitivity S right there where the TE_0 mode shows a sensitivity peak and no other mode exists, losing the possibility to have access to an independent identification of refractive index n and thickness of the adlayer d, or (2) choose a compromise, the highest combination of sensitivities with two modes, e.g., where the TE_1- and TM_1-mode sensitivity lines are crossing, allowing for n and d determination, but losing about a third in sensitivity. The decision has to be governed by the application of the device: if one needs to detect with the highest sensitivity whether a specific molecule is present in a solution and builds up an adlayer or not, sensitivity is the attribute to go for. If a new surface binding strategy is developed, where surface information such as the bonding density and final thickness of the adlayer are required, the compromise solution with two waveguide modes is adequate. Concentration information can also be found in the high-sensitivity case by calibrating the device, which basically serves as the second measurement necessary to gain information about a second variable.

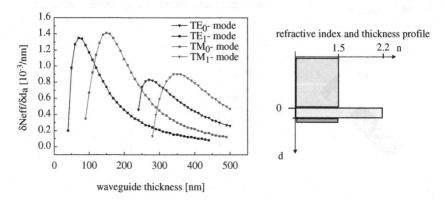

FIGURE 2.6. The sensitivity $S = \delta N/\delta d$ in 10^{-3} per nm adlayer is plotted for the first two TE and TM modes with respect to the thickness of a waveguide fabricated out of Ta_2O_5 having a refractive index of 2.2 at 632.8 nm. The adlayer is modeled with a refractive index of 1.5. The inset shows the waveguide architecture with respect to thicknesses and refractive indices.

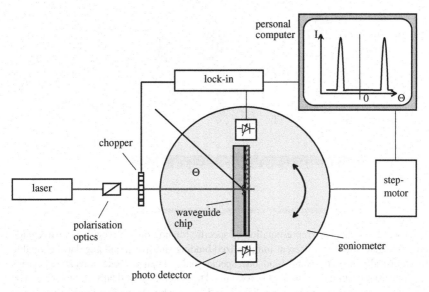

FIGURE 2.7. Experimental setup for the determination of the coupling angles.

Figure 2.7 shows the principle for a grating coupling measurement. The waveguide sample carrying a coupling grating is rotated on a goniometer, systematically changing the angle of incidence from a chopped laser with adjusted polarization. Detectors at the ends of the waveguide structure detect the modes as peaks, when light is coupled and guided to them. For sensor purposes the waveguide is connected to a cuvette with a liquid or gas handling system allowing liquids or gases to reach the waveguide surface and build up an adlayer. A lock-in technique is used to maximize the signal-to-noise-ratio.

2.3 Surface Functionalization and Reaction Recognition

When a sensor is needed, e.g., to identify whether a specific molecule is present in a solution or not, one wants to be sure that only the molecule under consideration is checked for and no similar molecules are involved in the same measurement. In other words a sensor needs to be specific.

How is it possible to ensure that the buildup of the above discussed adlayers is specific for only one molecular species? This is typically done by functionalizing the surface with a molecule which allows the molecule under consideration to bind specifically, via a recognition reaction or a guest–host reaction to it [39–42]. All other molecules should not bind to the immobilized matrix, avoiding unspecific binding. Figure 2.8 shows a schematic drawing of a waveguide carrying two different immobilized binding sites (oval and angular) to specifically bind only one fitting molecule each. Typically a waveguide carries only one species of binding site.

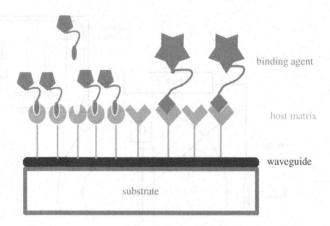

FIGURE 2.8. Schematic representation of specific binding on a waveguide: a waveguide on a substrate with two different immobilized binding sites (oval and angular) being able to specifically recognize and bind one species only. Typically one waveguide carries one functionalization only to avoid crosstalk. The scheme is vertically not to scale. The waveguide dimensions are between 200 nm and 1 μm, whereas the first immobilized layer has a thickness on the order of a few angstroms to a few nanometers.

Various chemical strategies are present to immobilize recognition sites on surfaces from Langmuir-Blodgett films to self-assembled monolayers [43–48]. A covalent link to the substrate enhances the stability of the functionalization on the surface, when liquids are rinsed across it. It depends on the kind of materials which are supposed to be connected. Glasses having -OH groups present at the surface after cleaning can be functionalized by a process called silanization. Here a covalent bond is formed between the oxygen of the surface with a Si from the immobilized silane [R(with function)- Si-R_3] undergoing several intermediate reaction states [39, 40]. Metal surfaces are easily functionalized via self-assembled monolayers (SAMs) formed out of molecules having a thiol, a sulfide, or a disulfide function in a spacer group [39, 43–47].

Sometimes multistep functionalizations are necessary: e.g., first a silanization which allows use of the immobilized silane function R to continue immobilizing the desired biological or chemical function [40, 48]. Multistep functionalization strategies cover established biological guest–host systems such as the biotin–avidin system [49–52] (Figure 2.9), but also approaches of metal nanoparticles [53, 54] have been made.

But whatever the choice of the chemistry adequate to link the "catcher" molecule to a surface, one has to make sure that the active binding site is accessible and that no steric hindrance occurs during the binding caused by the geometry of the catcher, the recognized molecule and the bound system. Various authors have shown that steric hindrance can be easily overcome by mixing the active catcher molecule with an inactive dilutor molecule (Figure 2.9). In most cases, when no phase separation in this binary mixture occurs [55–57], this decouples the binding sites laterally from each other and then provides easy

a

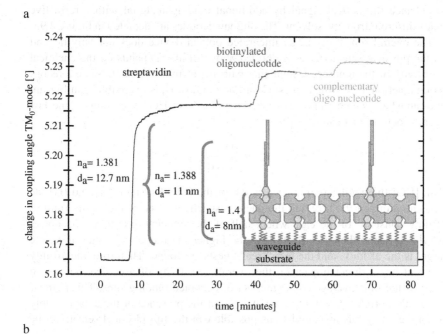

b

-Biotin-(T)30 **ACG TCA GTC TCA CCC**-3'
   3'-**TGC AGT CAG AGT GGG** TGT TAT AAG TGA GTC GGT GAT GAT GGA GAC ATT GA-5'

FIGURE 2.9. (a) Response of the coupling angle with time for a three-step surface functionalization scheme and an oligonucleotide hybridization reaction. The inset describes the layer system: a biotinylated thiol layer is immobilized on the waveguide, followed by streptavidin, a biotinylated oligonucleotide as catcher, and the complementary oligonucleotide for hybridization of the molecule under consideration to be detected. The given data are the overall layer thicknesses and refractive indices of the multilayer systems. (b) The oligonucleotide sequences.

accessibility [58–61]. If a recognition reaction involves a large interdigitation of the two molecules (e.g., for DNA hybridization) or binding into a pocket (e.g., biotin binds into a binding pocket in avidin), the molecules have to be located on a flexible spacer to overcome steric hindrance along the binding direction.

The inset of Figure 2.9a explains most of the above-mentioned characteristics for efficient multistep functionalization. On top of the waveguide a mixture of biotinylated silanes with a long alkyl chain mixed with inactive OH-terminated short-chain alkyl silanes is immobilized. Then a layer of streptavidin is bound to the biotin functionalization [62, 63]. In the next step, a biotinylated oligonucleotide [64–67] is bound to the steptavidin, exposing binding sites for biotin at its new surface. In this case, the synthetic oligonucleotide consists of 30 T (thymine) units as a spacer and 5 codons as the catcher [64–67]. The recognized molecule consists of the 5 matching codons and 12 additional random codons to enhance the optical signal by additional binding material with a refractive index different from the solvent. The oligonucleotides are depicted in Figure 2.9b.

The overall sensitivity of an integrated optical device does not only depend on the physical sensitivity S as defined above. It also depends on the chemical sensitivity of the binding, recognizing pair: the affinity constant. Since a recognition reaction, e.g., from a guest G into or to a host H, is reversible, equilibrium is established, which can be described as: $H + G \leftrightarrow HG$. According to the law of mass action, this can be written as

$$K = \frac{[HG]}{[H][G]}$$

with [HG] the equilibrium concentration of the bound system, [H] and [G] the concentrations of the host and the guest, respectively, and K the equilibrium or affinity constant. In the case where half of the available guests are occupied, $[HG] = [H]$ and then $K = 1/[G]$. So the higher the numerical value of K, the larger is the affinity, and the less "free" guests are found. Therefore, the affinity constant has a direct impact on the overall sensitivity of a device, because it controls the time average of the number of occupied binding sites. This number is directly correlated with the time average of the presence of the adlayer. Only when the adlayer is as complete as possible can the full physical sensitivity be achieved.

The operation of a sensor does not necessarily mean only the detection of molecules binding to a recognition site. It can also mean a detection of changes within the adlayer which lead to alterations of the thickness and/or the refractive index. These changes can have various causes including photochemical reactions, swelling, temperature effects such as transitions, aging, and wear.

The evanescent field can also be used to illuminate ultrathin samples located on top of a waveguide with the two specific polarization directions. This leads to two main applications: evanescent absorption spectroscopy [69–73] and evanescent microscopy [74–79]. A huge number of integrated optical sensors use the evanescent field to excite fluorescence labels attached to the molecules under consideration when they bind to their recognition site [80, 81, 82]. This chapter does not cover these very sensitive (bio) chemical sensors; rather, it covers only label-free methods. It also does not include all the studies in which surface plasmon resonances of gold dielectric interfaces on a waveguide are coupled to waveguide modes [83–85].

2.4 Examples

2.4.1 Monitoring of a Photochemical Isomerization Reaction in a Waveguide 1.1 µm Thick

Mesoionic 6-oxo-1,6-dihydropyrimidin-3-ium-4-oleate (Figure 2.10a) was incorporated as a part of a side chain into a polymer. The mesoion is connected to the

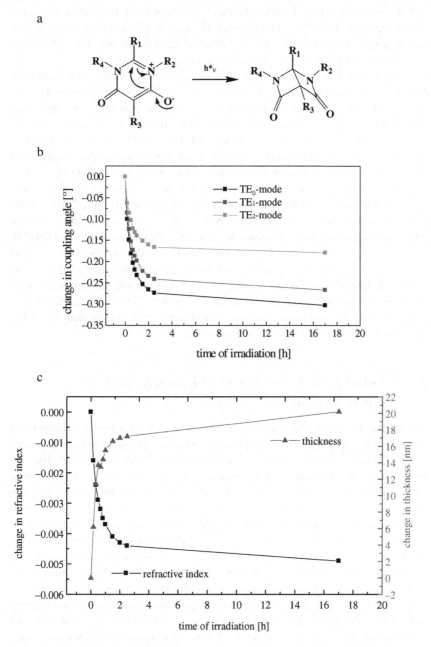

FIGURE 2.10. (a) Photoreaction of mesoionic 6-oxo-1,6-dihydropyrimidin-3-ium-4-oleate. (b) The change of three TE modes with illumination time. (c) Thickness and refractive index change with illumination time calculated from (b).

polymer main chain via a spacer group R_1. The rests R_2 and R_4 are benzene rings and R_3 is a CH_3 group. Further chemical details and synthesis strategies have been given elsewhere [86]. The mesoion changes its planar mesoionic structure into an angled bis(β-lactam) structure on irradiation in the near-UV. This leads to a decrease in polarizability and to a higher steric demand.

The polymer was spin-coated as a waveguide onto a fused silica substrate carrying a previously fabricated coupling grating [87] with a periodicity $\Lambda = 797$ nm. A HeNe laser at a wavelength of 632.8 nm was used to couple into waveguide modes at both polarization directions. In order to photoisomerize the polymer, a linearly polarized beam from a HeCd laser (442 nm) was incident at $0°$ to the coupling spot of the waveguide modes. Three modes of each polarization were found and followed with illumination time. Figure 2.10a shows the change in coupling angles versus illumination time for the TE modes. For all data sets the refractive index and thickness values were calculated (Figure 2.10c). Both polarization directions gave identical results; thus, a homogeneous refractive index and refractive index decrease were found. The diminished polarizability of the mesoion and the swelling of the film due to the higher steric demand after photobleaching are responsible for the refractive index drop. The increase in film thickness is direct evidence for the photoreaction and the enhancement of the necessary space for the photo-switched mesoion. This example shows how a thick film in the micrometer regime, and changes in it, can be analyzed using waveguide spectroscopy on a known substrate.

2.4.2 Label-Free Detection of Oligonucleotide Hybridization

As discussed in Section 2.3, Figure 2.9 shows a multilayer functionalization on a waveguide using the biotin–streptavidin system. The figure exhibits changes in the TE mode coupling angles of a commercial Ta_2O_5 waveguide (Unaxis) with a thickness of 160 nm and a coupling grating of $\Lambda = 360$ nm. The coupling angle changes on rinsing a new solution along the waveguide and binding the next species onto it. The first step is streptavidin adsorption, the second step the binding of the catcher oligonucleotide, and the third step the hybridization reaction of the oligonucleotides. From both the TE and the TM mode measurements the refractive index and the thickness of the entire adlayers on the waveguide were calculated for all steps. Streptavidin on the biotin–silane layer yields a refractive index of 1.4 and an overall thickness of 8 nm. This is a typical data set found for a water-swollen silane–protein system [88]. Immobilizing the catcher oligonucteotide leads to an increase in thickness by 3 nm to 11 nm and a decrease in refractive index to 1.388. This decrease in refractive index is due to a non-densely packed film: water is located in-between the oligonucleotide molecules. The final hybridization reaction leads to a further decrease in refractive index to 1.381, and an increase in thickness to 12.7 nm. Again the refractive index drop shows the water involved in the films deposited.

Taking DNA data into account [64, 65], a densely packed 45-mer would lead to an estimated film thickness increase of 15.3 nm. Obviously this is not seen by waveguide spectroscopy. The films are not densely packed and probably not even ordered, showing only elongated chains sticking out perpendicularly as drawn in Figure 2.9a. The same is true for the hybridization reaction with an additional 35-mer. The interpretation of such data has to be done with great care. Only one TE and one TM mode can be measured at that high sensitivity (see Figure 2.6). The data evaluation for thickness and refractive index is then done under the assumption of an isotropic film. The film here is not completely isotropic because of its layer structure. Therefore, only the overall data are calculated and neither single layer data nor any concentration data are given. The data are estimations, only showing trends. Additional information is necessary, such as the use of two more waveguide modes by choosing a thicker waveguide and losing sensitivity (Figure 2.6), or by a calibration measurement.

This example shows how a high-sensitivity waveguide can detect the binding of even minute amounts of molecules in a yes–no answer fashion: if a shift in the coupling angle is measured, the molecule under consideration is present (if the recognition reaction is specific). If no shift is seen, no binding occurred; the molecule was not present or only present in a concentration below the detection limit. Any other information, e.g. mismatches in the oligonucleotide hybridizing, has to be found by additional investigations [89–93].

2.4.3 Swelling of Thin Polymer Films in Solvent

An ultrathin film of polyallylamine with a thickness of 7.7 nm was deposited by a plasma polymerization reaction onto a commercial Ta_2O_5 waveguide (Unaxis) with a thickness of 160 nm and a coupling grating of $\Lambda = 360$ nm. The details of the film formation are published elsewhere [94–96]. The film was then placed at room temperature into ethanol. Ethanol is a good solvent for polyallylamine and should be able to intrude into the film and swell it. The coupling angles for both polarization directions were measured with increasing time in ethanol. Then the sample was dried in an oven. After cooling back to room temperature, the coupling angles were followed in time in regular air. Figure 2.11 shows the thickness (a) and the refractive index (b) data calculated from the optical waveguide data. The film was found to be isotropic. The film thickness increases over hours when located in ethanol. Even after 18 hours no equilibrium plateau was reached, but an increase in film thickness from 7.7 nm to 11 nm. The thorough drying of the film leads to an even smaller film thickness in comparison to the newly prepared film. This means that during the drying process a small amount of molecular material must have been removed, which was present in the newly prepared films. These are probably solvent leftovers from the cleaning procedure. A small increase in film thickness is detected in air. Obviously the film takes up gases from the air.

The refractive index, on the other hand, decreases with ethanol uptake. This is due to a dilution process. As can be seen from Figure 2.11b, the refractive

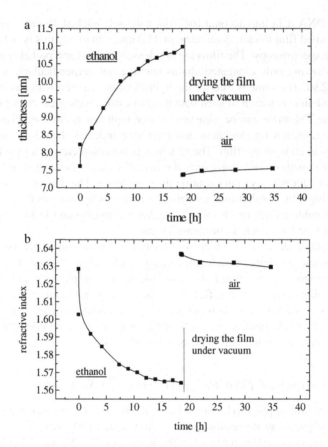

FIGURE 2.11. (a) Thickness and (b) refractive index behavior of a plasma-deposited polyallylamine film on swelling in ethanol, drying, and placement in air.

index of the dried, mostly pure material is 1.636. The intake of ethanol having a refractive index of 1.361 dilutes the high refractive index of the polymer and lets it decrease with an increasing amount of the low-index solvent. It can be recovered by drying the film and slowly decreases again during uptake of air gases.

This example shows how a waveguide can be used to monitor swelling effects due to chemical uptake in thin films within the nanometer regime.

2.4.4 Glass Transition Measurements in Polymer Mono- and Multi-layer LB Films

The glass transition T_g (freezing of segmental movement) in thin polymer films has attracted the interest of polymer researchers for quite some time, because it has been unclear whether a thin film material would behave differently from the bulk [97–111]. Waveguide attempts were very successful, here using "thick" waveguides and measuring the glass transition in the waveguide material itself [105–107]. The glass transition typically appears as a kink in the temperature–waveguide mode position, because at the glass transition the thermal expansion coefficient $\beta = \frac{1}{d}\frac{\delta d}{\delta T}$ (with d the thickness and T the temperature) changes as well as the thermal refractive index coefficient $k = \frac{\delta n}{\delta T}$ [106]. It was found that indeed the glass transition decreases with decreasing film thickness, as long as no interaction between the substrate and the polymer is present. Free standing films were also investigated with ellipsometry [90]. The smallest thickness films were prepared by LB technology [43–48] and investigated by surface plasmon resonance spectroscopy, an evanescent technology similar to waveguide spectroscopy, but with less information gain [106, 112].

Figure 2.12a shows the coupling angle behavior of the TE_0 mode of a commercial Ta_2O_5 waveguide (Unaxis) with a thickness of 160 nm and a coupling grating of $\Lambda = 360$ nm for $\lambda = 632.8$ nm with temperature changing (both increasing and decreasing). The coupling angle position increases with

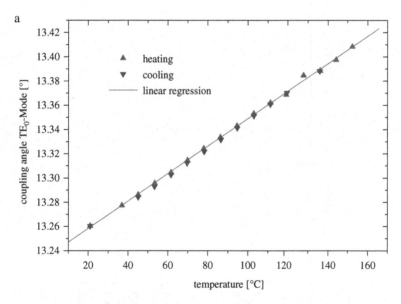

FIGURE 2.12. (a) Coupling angle behavior of the TE_0 mode of a Ta_2O_5 waveguide with temperature change: red data represent the heating, blue data the cooling. (b) The structure formula of the polyamide acid and the reaction into the polyamide releasing water. Two derivatives with different spacer lengths $n = 4$ and $n = 6$ were investigated. (c) Position of the coupling angle of the TM_0 mode of the Ta_2O_5 waveguide with one monolayer of the $n = 6$ spacer polymer in the polyamide form after heating with decreasing temperature. The lines are linear regressions to the upper (blue) or lower (red) data points, respectively. The temperature where both regressions intersect is the glass transition temperature of that monolayer.

FIGURE 2.12. (*Continued*)

c

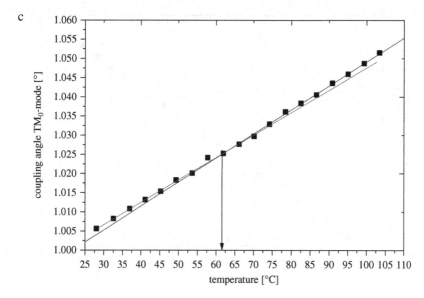

FIGURE 2.12. *(Continued)*

increasing temperature due to the thermal expansion β and the temperature refractive index coefficient κ of the waveguide sample itself. No hysteresis is observed. The data showed an identical behavior for the TM mode. This proved that the Ta_2O_5 waveguide is an excellent substrate for materials which show temperature-driven transitions.

Using the LB technique, monolayers and multilayers of polyamide acid can be transferred in the condensed state to Ta_2O_5 waveguides and subsequently heated to form polyimide under water production (Figure 2.12b). The bulk glass transition temperatures T_g for the two derivatives were determined via DSC investigations [113]. The polyimide with a spacer group length of $n = 4$ shows a T_g of $\sim 88\,°C$ and the $n = 6$ spacer polymer a T_g of $\sim 58°\,C$.

A single monolayer of the $n = 6$ material was transferred to the waveguide and converted to the polyamide, then investigated with respect to the TM waveguide mode position with decreasing temperature. Figure 2.12c shows the waveguide mode coupling data together with linear regressions to the upper (red line) and lower (blue line) temperature data. Clearly two different slopes are found. The intersection of the two lines, the kink, indicates a glass transition temperature T_g at $\sim 62°\,C$ very close to the bulk value at $\sim 58°\,C$. Because the glass transition is not a very sharp transition, these two values are identical within experimental error. Contrary to most of the literature, no decrease in the glass transition due to the ultrathin film geometry was found, probably because interactions between the substrate and polymer film were not inhibited. Because the waveguide setup allows for anisotropy investigations, multilayer samples were investigated with the two polarization directions. Figure 2.13a shows a nine-layer sample of the $n = 6$ spacer material investigated with the TM mode, whereas Figure 2.13b

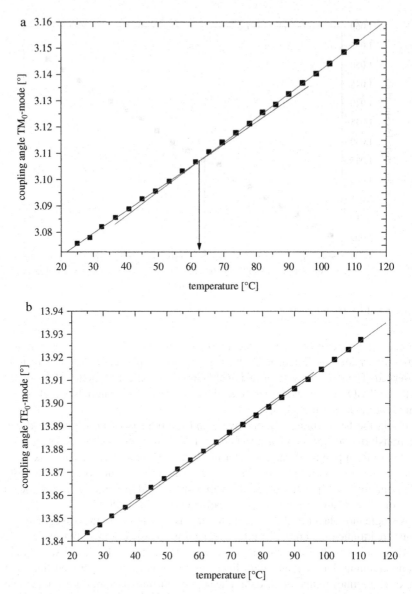

FIGURE 2.13. (a) Coupling angle behavior of the TM_0 mode for a nine layer sample of the $n = 6$ spacer material and (b) coupling angle behavior for a nine layer sample of the $n = 4$ spacer material.

shows data for a nine layer sample of the $n = 4$ material with the TE mode. Not surprisingly, the TM mode is able to show the same glass transition temperature for the thicker sample, but surprisingly the TE mode does not detect any kink with two different slopes in the temperature data, but an offset at around 70° C, which is a little below the T_g value of the $n = 4$ bulk material ($\sim 88°\mathrm{C}$). The physicochemical reason for this behavior is not yet well understood and it is still under investigation. The cause of this anisotropic behavior lies probably within the anisotropic structure of the monolayers fabricated by LB technology. The hydrophilic main chain of the polymer is "separated" by hydrophobic interactions from the hydrophobic side chains on the air–water interface, leading to a transferred first monolayer which is attached with the main chain to the hydrophilic substrate with side chains randomly sticking out. This looks like a nanoscopic comb. The TM mode then preferentially probes the side chains carrying the spacers, which are partly responsible for the precise position of T_g. The TE mode preferentially probes the main chain, which seems to only indirectly couple to T_g, showing an offset instead of a kink involving changes in physical constants.

This example exhibits how waveguide spectroscopy can be used for materials research down to monolayer sensitivity including anisotropic effects which are not seen with any other method at these sample thicknesses.

2.4.5 Evanescent Scattering Microscopy

As mentioned above, the evanescent field of a waveguide can also be used to excite molecules, typically in technologies involving fluorescence labels. On the other hand, one finds, looking at the mechanisms leading to waveguide losses, that with waveguides having surfaces that have not been prepared smoothly, light escapes — it is scattered by surface inhomogeneities. This effect is used in evanescent scattering microscopy [74, 75]. Here a laterally structured thin film sample is placed on a waveguide into the evanescent field of the propagating modes and the scattered light is observed with a standard microscope perpendicular under 0° angle distortion free. Figure 2.14 shows a typical setup. A polarized laser beam is coupled by a grating coupler to a specific mode (polarization and mode number). A cuvette enables microscopy to be performed in either a solvent or an aqueous environment. Figure 2.15 shows images (a) with the TE_3 mode and (b) with the TM_3 mode, taken from a four-mode

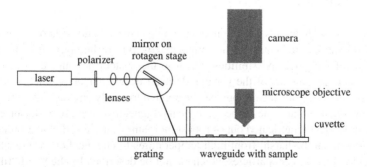

FIGURE 2.14. Experimental setup for evanescent waveguide scattering microscopy.

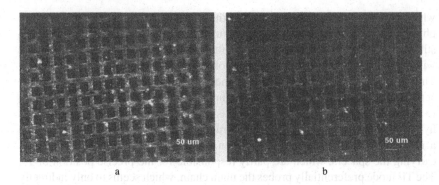

FIGURE 2.15. Scattering microscopy image taken of a 40-nm-thick laterally patterned photoresist sample on illuminating with (a) the TE$_3$ mode and (b) the TM$_3$ mode.

ion exchanged waveguide in BK7 glass with a 40-nm-thick laterally patterned photoresist sample. The photoresist appears bright in the images, because it creates inhomogeneities in the evanescent field which lead to scattered photons giving a bright impression. The dark areas are the pure waveguide surfaces which do not scatter and are therefore dark. The TE mode shows a higher contrast than the TM mode. It was also found that with increasing mode number, for both polarization directions the contrast increased. Simulations of the evanescent field strength have demonstrated that the contrast is directly correlated with the present strength of the evanescent field occupied by the scattering film. In addition, the contrast is a function of the refractive index contrast between the scattering material and the surrounding material. A photoresist–air image shows much higher contrast than a photoresist–water image. Because a conventional microscope is used, the lateral resolution is limited by diffraction and is therefore in the micrometer regime. The resolution in the z-direction is at least 40 nm, as measured. The minimal thickness still being seen with decent contrast has not yet been determined.

2.5 Conclusion

As we have seen in this selection of examples, waveguide spectroscopy and especially the evanescent field accompanied by this technology offer a wide spectrum of analytical possibilities. Because the system is coupled to at least two physical parameters of the waveguide itself or an adlayer — thickness and refractive index — this technology can be applied to nearly any thin film analysis problem, where at least one of these physical parameters is modified. Because the waveguides can be fabricated out of robust inorganic materials, the measurement environment can be harsh: from high temperatures to chemical environments.

Samples such as nanoparticles which cannot be described by the word "film," meaning that there is no lateral dimension which is "infinite" in parallel direction to the waveguide surface, cannot be analyzed by this technology so far, because the theoretical background for data analysis is based in the Fresnel equations — which in their classical form are valid only for layer systems with homogeneous thicknesses over large dimensions. Nevertheless, by additionally taking into account other light-guiding systems such as surface plasmon spectroscopy, and waveguide architectures and applications (from waveguides with metal coatings, surface plasmon waveguide mode coupling, through to devices based on fluorescence to waveguide absorption spectroscopy and microscopy technologies), we see an amazing spectrum of nanoscale analytical possibilities.

Further research and development, combining nanostructures such as metal nanoparticles or photonic crystals with conventional waveguide geometries and architectures, will lead to new physical effects and definitely to new nano-optical devices for applications from data storage to all-optical switching, communications technology, and, again, the analysis of nanostructures.

Acknowledgments. The authors thank all co-workers and colleagues who have contributed over the years in using waveguides and their fields as analytical tools.

References

1. *Bergmann-Schaefer, Lehrbuch der Experimentalphysik, Band III Optik*, de Gruyter, Berlin (1978).
2. C. Gerthsen, H.O. Kneser, and H. Vogel, *Physik*, 13th edition, Springer-Verlag, Berlin (1977).
3. J.D. Jackson, *Classical Electrodynamics*, 3rd edition, Wiley, New York (1999).
4. M.J. Adams, *An Introduction to Optical Waveguides*, Wiley, New York (1981).
5. R.G. Hunsperger, *Integrated Optics: Theory and Technology*, 4th edition, Springer-Verlag, Berlin (1995).
6. T. Tamir (Ed.), *Integrated Optics*, 2nd edition, Springer-Verlag, Berlin (1979).
7. A.W. Snyder and J.D. Love, *Optical Waveguide Theory*, Chapman & Hall, London (1983).
8. R. Syms and J. Cozens, *Optical Guided Waves and Devices*, McGraw–Hill, New York (1992).
9. D. Marcuse, *Theory of Dielectric Optical Waveguides*, 2nd edition, Academic Press, Boston (1991).
10. G.P. Agrawal, *Applications of Nonlinear Fiber Optics*, Academic Press, San Diego (2001).
11. C.D. Chaffee, *Building the Global Fiber Optics Superhighway*, Kluwer Academic/ Plenum, New York (2001).
12. G.D. Cole, *Computer Networking for Systems Programmers*, Wiley, New York (1990).
13. R. Ulrich and R. Torge, Measurement of thin film parameters with a prism coupler, *Appl. Opt.* **12**, 2901–2908 (1973).
14. W. Karthe and R. Müller, *Integrierte Optik*, Akademische Verlagsgesellschaft Geest & Portig, Leipzig (1991).

15. T. Liu and R.J. Samuels, Physically corrected theoretical prism waveguide coupler model, *J. Opt. Soc. Am. A* **21**, 1322–1333 (2004).
16. M.A. Uddin, H.P. Chan, and C.K. Chow, Thermal and chemical stability of a spin-coated epoxy adhesive for the fabrication of a polymer optical waveguide, *Chem. Mater.* **16**, 4806–4811 (2004).
17. X. Zhang, H.J. Lu, A.M. Soutar, and X.T. Zeng, Thick UV-patternable hybrid sol-gel films prepared by spin coating, *J. Mater. Chem.* **14**, 357–361 (2004).
18. D.A. Chang-Yen and B.K. Gale, An integrated optical oxygen sensor fabricated using rapid-prototyping techniques, *Lab on a Chip* **3**, 297–301 (2003).
19. G.M. Wang, H. Ai, X. Liu, and F.G. Tao, A nonlinear optical waveguide made up of alternating NMOB/CdA multilayers, *J. Opt. A Pure Appl. Opt.* **4**, 587–590 (2002).
20. M. Yanagida, A. Takahara, and T. Kajiyama, Construction of defect-diminished fatty acid Langmuir-Blodgett film and its optical waveguide properties, *Bull. Chem. Soc. Jpn.* **72**, 2795–2802 (1999).
21. W. Hickel, G. Appel, D. Lupo, W. Prass, and U. Scheunemann, Langmuir-Blodgett multilayers from polymers for low-loss planar waveguides, *Thin Solid Films* **210**, 182–184 (1992).
22. C. Bosshard, M. Florsheimer, M. Kupfer, and P. Günter, Cerenkov-type phase-matched 2nd- harmonic generation in DCANP Langmuir-Blodgett-film waveguides, *Opt. Commun.* **85**, 247–253 (1991).
23. S. Mittler-Neher, A. Otomo, G.I. Stegeman, C. Y.-C. Lee, R. Mehta, A.K. Agrawal, and S.A. Jenekhe, Waveguiding in substrate and freestanding films of insoluble polymers, *Appl. Phys. Lett.* **62**, 115–117 (1993).
24. G. Ihlein, B. Menges, S. Mittler-Neher, J.A. Osaheni, and S.A. Jenekhe, Channel waveguides of insoluble conjugated polymers, *Opt. Mater.* **4**, 685–689 (1995).
25. C.C. Huang, D.W. Hewak, and J.V. Badding, Deposition and characterization of germanium sulphide glass planar waveguides, *Opt. Express* **12**, 2501–2506 (2004).
26. N.J. Goddard, K. Singh, J.P. Hulme, C. Malins, and R.J. Holmes, Internally-referenced resonant mirror devices for dispersion compensation in chemical sensing and biosensing applications, *Sens. Actuat. A* **100**, 1–9 (2002).
27. D.P. Bour, M. Kneissl, C.G. Van de Walle, G.A. Evans, L.T. Romano, J. Northrup, M. Teepe, R. Wood, T. Schmidt, S. Schoffberger, and N.M. Johnson, Design and performance of asymmetric waveguide nitride laser diodes, *IEEE J. Quantum Electron.* **36**, 184–191 (2000).
28. Q. Gao, M. Buda, H.H. Tan, and C. Jagadish, Room-temperature preperation of InGaAsN quantum dot lasers grown by MOCVD, *Electrochem. Solid State Lett.* **8**, G57–G59 (2005).
29. M. Born and E. Wolf, *Principles of Optics*, Pergamon Press, London (1970).
30. S.I. Najafi, *Introduction to Glass Integrated Optics*, Artech House, Boston (1992).
31. R.V. Ramaswamy and R. Srivastavam, Ion-exchanged glass waveguides — a review, *J. Lightwave Technol.* **6**, 984–1002 (1988).
32. P.D. Townsend, P.J. Chandler, and L. Zhang, *Optical Effects of Ion Implantation*, Cambridge University Press, Cambridge (1994).
33. S. Mittler-Neher and H. Einsiedel, Photothermal beam deflection techniques: a useful tool for integrated optics, *Proc. SPIE* **2852**, 248–257 (1996).
34. S. Mittler-Neher, Linear optical characterization of nonlinear optical polymers for integrated optics, *Macromol. Chem. Phys.* **199**, 513–523 (1998).

35. H. Einsiedel, M. Kreiter, M. Leclerc, and S. Mittler-Neher, Photothermal beam deflection spectroscopy in the near IR on poly[3-alkylthiophene]s, *Opt. Mater.* **10**, 61–68 (1998).

36. G. Lenz and J. Salzman, Eigenmodes of multiwaveguide structures, *J. Lightwave Technol.* **8**, 1803–1809 (1990).

37. W. Lukos and K. Tiefenthaler, Sensitivity of integrated optical grating and prism couplers as (bio)chemical sensors, *Sensors Actuat.* **15**, 273–284 (1988).

38. K. Tiefenthaler and W. Lukosz, Sensitivity of grating couplers as integrated-optical chemical sensors, *J. Opt. Soc. Am. B* **6**, 209–220 (1989).

39. P. Mohr, M. Holtzhauer, and G. Kaiser, *Immunosorption Techniques: Fundamentals and Applications*, Akademie-Verlag, Berlin (1992).

40. T. Cass and F.S. Ligler (Eds.), *Immobilized Biomolecules in Analysis*, Oxford University Press, Oxford (1998).

41. E.M. Blalock (Ed.), *A Beginner's Guide to Microarrays*, Kluwer Academic, Boston (2003).

42. D. Wild (Ed.), *The Immunoassay Handbook*, 2nd edition, Nature Publishing Group, New York (2001).

43. A. Ulman, *An Introduction to Ultrathin Organic Films: From Langmuir-Blodgett to Self-Assembly*, Academic Press, Boston (1991).

44. A. Ulman, Formation and structure of self-assembled monolayers, *Chem. Rev.* **96**, 1533–1554 (1996).

45. A. Ulman, *Organic Thin Films and Surfaces: Directions for the Nineties*, Academic Press, San Diego (1995).

46. A. Ulman (Ed.), *Characterization of Organic Thin Films*, Butterworth-Heinemann, Boston (1995).

47. F. Schreiber, Structure and growth of self-assembling monolayers, *Prog. Surf. Sci.* **65**, 151–256 (2000).

48. M.C. Petty, *Langmuir-Blodgett Films: An Introduction*, Cambridge University Press; Cambridge (1996).

49. M. Wilchek and E.A. Bayer, *Avidin-Biotin Technology*, Academic Press, San Diego (1990).

50. V.C. Yang and T. T. Ngo (Eds.), *Biosensors and Their Applications*, Kluwer Academic/Plenum, New York (2000).

51. D.Möbius and R. Miller (Eds.), *Organized Monolayers and Assemblies: Structure, Processes, and Function*, Elsevier, Boston (2002).

52. C.M. Niemeyer, *Bioconjugation Protocols: Strategies and Methods*, Humana Press, Totowa, NJ (2004).

53. S. Busse, J. Käshammer, S. Krämer, and S. Mittler-Neher, Gold and thiol surface functionalized integrated optical Mach-Zehnder interferometer for sensing purposes, *Sens. Actuat. B* **60**, 148–154 (1999).

54. Z.-M. Qi, N. Matsuda, T. Yoshida, A. Takatsu, and K. Kato, Colloidal gold submonolayer- coated thin-film glass plates for waveguide-coupled surface plasmon resonance sensors, *Appl. Opt.* **42**, 4522–4528 (2003).

55. N.J. Brewer and G.L. Leggett, Chemical force microscopy of mixed self-assembled monolayers of alkanethiols on gold: Evidence for phase separation, *Langmiur* **20**, 4109–4115 (2004).

56. F.Q. Fan, C. Maldarelli, and A. Couzis, Fabrication of surfaces with nanoislands of chemical functionality by the phase separation of self-assembling monolayers on silicon, *Langmuir* **19**, 3254–3265 (2003).

57. K. Aoki, Theory of phase separation of binary self-assembled films, *J. Electroanal. Chem.* **513**, 1–7 (2001).
58. J. Spinke, M. Liley, F.-J. Schmitt, H.-J. Guder, L. Angermaier, and W. Knoll, Molecular recognition at self-assembled monolayers: Optimization of surface functionalization, *J. Chem. Phys.* **99**, 7012–7019 (1993).
59. J. Spinke, M. Liley, H.-J. Guder, L. Angermaier, and W. Knoll, Molecular recognition at self-assembled monolayers: the construction of multicomponent multilayers, *Langmuir* **9**, 1821–1825 (1993).
60. M. Weisser, G. Nelles, G. Wenz, and S. Mittler-Neher, Guest–host interaction with immobilized cyclodextrins, *Sens. Actuat. B* **38–39**, 58–67 (1997).
61. M. Weisser, J. Käshammer, J. Matsumoto, F. Nakamura, K. Ijiro, M. Shimomura, and S. Mittler-Neher, Adenin–uridin base pairing at the water-solid-interface, *J. Am. Chem. Soc.* **122**, 87–95 (2000).
62. G. Tovar, L. Angermaier, P. Sluka, H.G. Batz, and W. Knoll, Molecular recognition at biotin-functionalized oxidic surfaces: Biofunctional self-assembly of proteins, *Abstr. Pap. Am. Chem. Soc.* **212**, 12 (1996) Part 1.
63. M. Weisser, G. Tovar, S. Mittler-Neher, W. Knoll, F. Brosinger, H. Freimuth, M. Lacher, and W. Ehrfeld, Specific bio-recognition reactions observed with an integrated Mach-Zehnder interferometer, *Biosens. Bioelectron.* **14**, 405–411 (1999).
64. A.L. Lehninger, *Principles of Biochemistry*, Worth Publishers, New York (1982).
65. D.A. Metzler, *Biochemistry: The Chemical Reactions of Living Cells*, Volume 1, 2nd edition, Harcourt Academic Press, San Diego (2001).
66. G. Stengel, F. Höök, and W. Knoll, Viscoelastic modeling of template-directed DNA synthesis, *Anal. Chem.* **77** (11)3709–3714(2005).
67. G. Stengel and W. Knoll, Surface plasmon field-enhanced fluorescence spectroscopy studies of primer extension of surface-bound oligonucleotides, *Nucleic Acids Res.* **33** (7)E69 (2005).
68. V. Ruddy, Nonlinearity of absorbency with sample concentration and path-length in evanescent wave spectroscopy using optical fiber sensors, *Opt. Eng.* **33**, 3891–3894 (1994).
69. K. Kato, A. Takatsu, N. Matsuda, R. Azumi, and M. Matsumoto, A slab waveguide absorption spectroscopy of Langmuir-Blodgett films with a white light excitation source, *Chem. Lett.* **6**, 437–438 (1995).
70. H. Kawai, K. Nakano, and T. Nagamura, White light optical waveguide detection of transient absorption spectra in ultrathin organic films upon pulsed laser excitation, *Chem. Lett.* **12**, 1300–1301 (2001).
71. J.H. Santos, N. Matsuda, Z.M. Qi, A. Takatsu, and K. Kato, Effect of surface hydrophilicity and solution chemistry on the adsorption behaviour of cytochrome c in quartz studied using slab optical waveguide spectroscopy, *IEEE Trans. Electron.* **E85-C**, 1275–1281 (2002).
72. J.T. Bradshaw, S.B. Mendes and S.S. Saavedra, A simplified broadband coupling approach applied to chemically robust sold-gel, planar integrated optical waveguides, *Anal. Chem.* **74**, 1751–1759 (2002).
73. F.T. Bradshaw, S.B. Mendes, N. Armstrong, and S.S. Saavedra, Broadband coupling into a single-mode, electroactive integrated optical waveguide for spectroelectrochemical analysis of surface-confined redox couples, *Anal. Chem.* **75**, 1080–1088 (2003).
74. F. Thoma, U. Langbein, and S. Mittler-Neher, Waveguide scattering microscopy, *Opt. Commun.* **134**, 16–20 (1997).

75. F. Thoma, J. Armitage, H. Trembley, B. Menges, U. Langbein, and S. Mittler-Neher, Waveguide scattering microscopy in air and water, *Proc. SPIE* **3414**, 242–249 (1998).

76. W. Knoll, W. Hickel, M. Sawodny, and J. Stumpe, Polymer interface and ultrathin films characterization by optical evanescent wave techniques, *Makromol. Chem. Makromol. Symp.* **48**, 363–379 (1991).

77. W. Knoll, W. Hickel, M. Sawodny, J. Stumpe, and H. Knobloch, Novel optical techniques for the analysis of polymer surfaces and thin films, *Fresenius J. Anal. Chem.* **341**, 272–278 (1991).

78. E.F. Aust and W. Knoll, Electrooptical waveguide microscopy, *J. Appl. Phys.* **73**, 2705–2708 (1993).

79. M. Bastmeyer, H.B. Deising, and C. Bechinger, Force exertion in fungal infection, *Annu. Rev. Biophys. Biomol. Struct.* **31**, 321–341 (2002).

80. X.F. Wang and U.J. Krull, Synthesis and fluorescence studies of thiazole orange tethered onto oligonucleotide: Development of a self-contained DNA biosensor on a fiber optic surface, *Bioorg. Med. Chem. Lett.* **15**, 1725–1729 (2005).

81. R.A. Yotter, L.A. Lee, and D.M. Wilson, Sensor technologies for monitoring metabolic activity in single cells — Part I: Optical methods, *IEEE Sens. J.* **4**, 395–411 (2004).

82. T.J. Pfefer, L.S. Matchette, A.M. Ross, and M.N. Ediger, Selective detection of fluorophore layers in turbid media: The role of fiber-optic probe design, *Opt. Lett.* **28**, 120–122 (2003).

83. M. Weisser, B. Menges, and S. Mittler-Neher, Multimode integrated optical sensor based on absorption due to resonantly coupled surface plasmons, *Proc. SPIE* **3414**, 250–256 (1998).

84. M. Weisser, B. Menges, and S. Mittler-Neher, Refractive index and thickness determination of monolayers by multi mode waveguide coupled surface plasmons, *Sens. Actuat. B* **56**, 189–197 (1999).

85. M. Weisser, J. Käshammer, J. Matsumoto, F. Nakamura, K. Ijiro, M. Shimomura, and S. Mittler-Neher, Adenin–uridin base pairing at the water-solid-interface, *J. Am. Chem. Soc.* **122**, 87–95 (2000).

86. A. Theis, B. Menges, S. Mittler, M. Mierzwa, T. Pakula, and H. Ritter, Polymeric mesoions: Novel synthetic methods, photochemistry, and characterization in the solid state by dielectric and waveguide-mode spectroscopy, *Macromolecules* **36**, 7520–7526 (2003).

87. X. Mai, R.S. Moshrefzahdeh, U.J. Gibson, G.I Stegeman, and C.T. Seaton, Simple versatile method for fabricating guided wave gratings, *Appl. Opt.* **24**, 3155–3161 (1985).

88. S. Busse, V. Scheumann, B. Menges, and S. Mittler, Sensitivity studies for specific binding reactions using the biotin/streptavidin system by evanescent optical methods, *Biosens. Bioelectron.* **17**, 704–710 (2002).

89. T. Liebermann, W. Knoll, P. Sluka, and R. Herrmann, Complement hybridization from solution to surface-attached probe-oligonucleotides observed by surface-plasmon-field enhanced fluorescence spectroscopy, *Colloids Surf. A* **169**, 337–350 (2000).

90. X.D. Su, R. Robelek, Y.J. Wu, G.Y. Wang, and W. Knoll, Detection of point mutation and insertion mutations in DNA using a quartz crystal microbalance and MutS, a mismatch binding protein, *Anal. Chem.* **76**, 489–494 (2004).

91. D. Kambhampati, P.E. Nielsen, and W. Knoll, Investigating the kinetics of DNA–DNA and PNA–DNA interactions using surface plasmon resonance-enhanced fluorescence spectroscopy, *Biosens. Bioelectron.* **16**, 1109–1118 (2001).

92. Y.T. Long, C.Z. Li, T.C. Sutherland, H.B. Kraatz, and J.S. Lee, DNA mismatch detection on gold surface by electrochemical impedance spectroscopy, *Abstr. Pap. Am. Chem. Soc.* 227: U88–U88 063-ANYL Part 1 (2004).

93. B. Dubertret, M. Calame, and A.J. Libchaber, Single-mismatch detection using gold-quenched fluorescent oligonucleotides, *Nature Biotechnol.* **19**, 365–370 (2001).

94. M.T. van Os, B. Menges, R. Förch, R.B. Timmons, G.J. Vancso, and W. Knoll, Thin film plasma deposition of allylamine effects of solvent treatment, *Mater. Res. Soc. Symp. Proc.* **544**, 45–50 (1999).

95. V. Jacobsen, B. Menges, R. Förch, S. Mittler, and W. Knoll, In-situ thin film diagnostics using waveguide mode spectroscopy, *Thin Solid Films* **409**, 185–193 (2002).

96. V. Jacobsen, B. Menges, A. Scheller, R. Förch, S. Mittler, and W. Knoll, In situ film diagnostics during plasma polymerization using waveguide mode spectroscopy, *Surf. Coat. Technol.* **142–144**, 1105–1108 (2001).

97. J.A. Forrest and K. Dalnoki-Veress, The glass transition in thin polymer films, *Adv. Colloid Interface Sci.* **94**, 167–196 (2001).

98. J.A. Forrest, K. Dalnoki-Veress, J.R. Stevens, and J.R. Dutcher, Effect of free surfaces on the glass transition temperature of thin polymer films, *Phys. Rev. Lett.* **77**, 2002 (1996).

99. J.A. Forrest, K. Dalnoki-Veress, and J.R. Dutcher, Interface and chain confinement effects on the glass transition temperature of thin polymer films, *Phys. Rev. E* **56**, 5705– 5716 (1997).

100. W.E. Wallace, J.H. van Zanten, and W.L. Wu, Influence of an impenetrable interface on a polymer glass transition temperature, *Phys. Rev. E* **52**, R3329–R3332 (1995).

101. Y. Grohens, M. Brogly, C. Labbe, M.-O. David, and J. Schultz, Glass transition of stereoregular poly(methyl methacrylate) at interfaces, *Langmuir* **14**, 2929–2932 (1998).

102. J.K. Keddie, R.A.L. Jones, and R.A. Cory, Size dependent depression on the glass transition temperature in polymer films, *Europhys. Lett.* **27**, 59 (1994).

103. S. Kawana and R.A.L. Jones, Character of the glass transition in thin supported films, *Phys. Rev. E* **63**, Art. No. 021501 (2001).

104. K. Dalnoki-Veress, J.A. Forrest, C.X. Murray, C. Gigault, and J.R. Dutcher, Molecular weight dependence of reductions in the glass transiton temperature of thin, freely standing polymer films, *Phys. Rev. E* **63**, Art.No. 031801 (2001).

105. H. Bock, S. Christian, W. Knoll, and J. Vydra, Determination of the glass transition temperature of nonlinear optical planar polymer waveguides by attenuated total reflection spectroscopy, *Appl. Phys. Lett.* **71**, 3643 (1997).

106. O. Prucker, S. Christian, H. Bock, J. Rühe, C.W. Frank, and W. Knoll, On the glass transition in ultrathin polymer films of different molecular architecture, *Macromol. Chem. Phys.* **199**, 1435–1444 (1998).

107. R. Kügler and W. Knoll, Influence of the polymethacrylate chain formation in Langmuir-Blodgett-Kuhn films on the glass transition temperature, *Macromol. Chem. Phys.* **203**, 923–930 (2002).

108. K. Dalnoki-Veress, J.A. Forrest, P.G. de Gennes, and J.R. Dutcher, Glass transition reductions in thin freely-standing polymer films: A scaling analysis of chain confinement effects, *J. Phys. IV* **10**, 221–226 (2000).

109. P.G. de Gennes, Glass transitions in thin polymer films, *Eur. Phys. J. E* **2**, 201–203 (2000).

110. G. Rehage and W. Borchard, in *The Physics of Glassy Polymers*, edited by R.N. Haward, Applied Intersciences, London (1973).

111. *Polymer Handbook*, 2nd edition, edited by J. Brandup and E.H. Immergut, Wiley, New York (1975).

112. H. Raether, *Surface Plasmons*, Springer, Berlin (1988).

113. G.W.H. Höhne, W.F. Hemminger, and H.-J. Flammersheim, *Differential Scanning Calorimetry*, Springer, Berlin (2003).

3

Porous Silicon Electrical and Optical Biosensors

Huimin Ouyang, Marie Archer,[1,2] and Philippe M. Fauchet[1]

Department of Electrical and Computer Engineering and Center for Future Health, University of Rochester, Rochester, NY 14627

3.1 Materials Science of Porous Silicon

Porous silicon (PSi) is a form of silicon with unique properties, distinct from those of crystalline, microcrystalline, or amorphous silicon. It was first prepared in 1956 [1] and much later it was identified as etched silicon [2]. The most common fabrication technique to produce PSi is electrochemical etching of a crystalline silicon wafer in a hydrofluoric (HF) acid-based solution [3]. The electrochemical process allows for precise control of the structural properties of PSi such as thickness of the porous layer, porosity, and average pore diameter. The morphology of PSi is important for sensing applications because the pore diameter limits the size of the species that can be captured.

PSi can be prepared with a wide range of optical and electrical properties which makes it a very flexible material. The internal surface of PSi is very large, ranging from a few to hundreds of square meters per gram [4]. Therefore, the properties of PSi are affected not only by its crystalline core and nanomorphology but also by its surface. In addition, the properties of the crystalline core differ from those of bulk c-Si, when the size of the silicon structures drops below 10 nm, a size regime in which quantum mechanical effects modify its electronic states [5, 6].

3.1.1 Control the Morphology and Porosity

The dissolution of silicon requires the presence of fluorine ions (F^-) and holes (h^+). The pore initiation and growth mechanisms are qualitatively understood.

---

[1] Also with the Department of Biomedical Engineering, University of Rochester.
[2] Permanent address: U.S. Naval Research Laboratory, Center for Biomolecular Science and Engineering, Washington, DC 20375.

Pore growth can be explained by several models, each one of which is more relevant in a specific regime of porosity and pore size [7–9]. If the silicon/electrolyte interface becomes irregular shortly after etching starts, the surface fluctuations of the Si/electrolyte interface may either grow (PSi formation) or disappear (electropolishing). A depletion region is formed in the thin PSi layer itself and also in a region of the Si wafer near the PSi/c-Si interface. In forward bias for p-type substrates, the holes can still reach the Si/electrolyte interface as the electric field lines are focused at the tip of the pores. Thus, holes will preferentially reach the Si/electrolyte interface deep in the pores, where etching will proceed rapidly (Figure 3.1). In contrast, no holes will reach the end of the Si rods, effectively stopping the etching there. In addition, if this electrostatic effect is not strong enough, the random walk of the holes toward the Si/electrolyte interface makes it more likely that they reach it at or near the pore's tip, resulting in an effect similar to the electrostatic one. When an n-type substrate is used, porous silicon formation takes place in reverse bias since holes are required for etching to proceed. Another important mechanism becomes predominant if the Si rods are narrow enough (typically much less than 10 nm). In this size regime, the electronic states start to differ from those of bulk silicon. When the motion of carriers is restricted in one or more dimensions, the holes in the valence band are pushed to lower energy by quantum confinement, which produces a potential barrier to hole transport from the wafer to the Si rods. The holes can no longer drift or diffuse into the Si rods and further etching stops.

Pore formation occurs when the fluorine ions are delivered faster than the holes, the interpore regions of PSi are depleted of holes, and further etching occurs only at the pore tips. When the current density decreases, the number of holes at the pore tips drops, which leads to smaller pore sizes. Thus, the

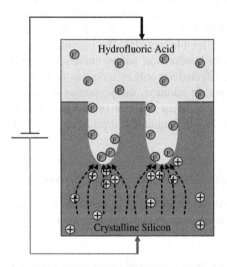

FIGURE 3.1. Schematic of PSi formation. Etching occurs only at the pore tips where the holes (h$^+$) are focused by the electric field.

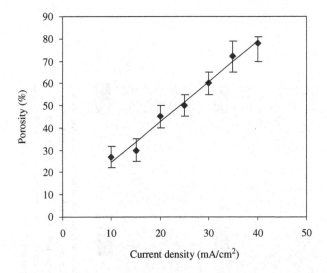

FIGURE 3.2. Porosity as a function of etching current density. The data are obtained from highly doped n-type silicon using 5% HF etching solution.

porosity (defined as the percentage of void space in the material) can be precisely controlled by the etching current density. Figure 3.2 shows the dependence of the porosity on current density for highly doped n-type (0.01 ohm-cm) silicon [10]. Similar curves can be found on other types of substrates [9].

The pore morphology is strongly affected by the choice of doping type and concentration, as illustrated in Figure 3.3 [11]. The top-row figures are top-view SEM images of porous silicon samples with different pore diameters ranging from mesopores (pore size >10 nm, < 50 nm) to macropores (pore size >50 nm). The bottom-row figures are cross sectional SEM images of the same samples. The mesopores formed in p+ silicon substrates have very branchy pore walls (Figure 3.3a, b). These pores only allow the detection of small objects such as short DNA segments and chemicals. The macropores formed in n-type wafers (Figure 3.3c–h) have much smoother pore walls and larger pore sizes than the mesopores. They are suitable for the detection of macromolecules and even very large objects such as bacteria or viruses. Note that most mesoporous silicon samples and certainly all microporous silicon samples (pore sizes < 10 nm) exhibit strong luminescence in the visible to near-infrared region [12]. This results from quantum confinement of electrons and holes in nanometer-sized quantum structures, which increases the band gap [5, 6] and enhances the radiative recombination rate [13].

3.1.2 Control of the Physical Properties

As already mentioned, the optical and electrical properties of PSi can be tuned by changing various parameters. The most spectacular effect is the tuning of the

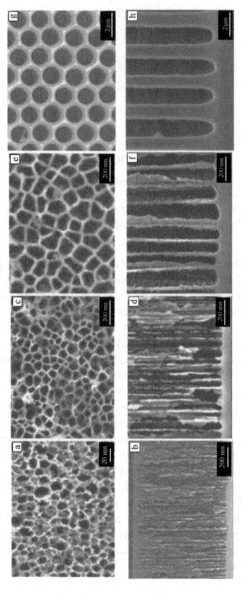

FIGURE 3.3. (a, b) Top-view and cross-sectional SEM images of mesoporous silicon with an average pore diameter of approximately 20 nm. It is formed in highly doped p-type silicon (0.01 ohm-cm) using an electrolyte with 15% HF in ethanol. (c, d) SEM images of 60 -nm macropores formed in very highly doped n-type silicon (0.001 ohm-cm) using an electrolyte with 6% HF. (e, f) SEM images of 120-nm macropores formed in highly doped n-type silicon (0.01 ohm-cm) using an electrolyte with 6% HF. (g, h) SEM images of 1.5-μm macropores etched from low-doped p-type silicon (20 ohm-cm) using an HF/dimethylformamide electrolyte.

luminescence from the band gap of bulk silicon to the UV by decreasing the size of the silicon remnants [12, 14]. The average size of the silicon remnants can be reduced to less than 2 nm by controlling the etching conditions (typically, by increasing the porosity) or by a postetching treatment (for example, oxidation). In this size regime, quantum-mechanical effects become important and modify the electronic states [13]. Quantum confinement increases the band gap and shifts the luminescence to shorter wavelengths, as shown in Figure 3.4 [14].

Immediately after etching, the internal surface of PSi is covered by silicon hydride bonds that produce an excellent surface passivation and, in the case of microporous silicon, strong luminescence. However, the luminescence of freshly etched PSi degrades quickly because the Si-H bonds can easily be broken, leading to dangling bonds that form nonradiative recombination centers [15]. It is thus useful to passivate the internal surface of PSi with SiO_2. This can be accomplished by various techniques, such as furnace oxidation or exposure to hydrogen peroxide. The stability of the luminescence efficiency of PSi passivated by SiO_2 has been shown to be excellent [16]. The development of biosensors using oxide-passivated PSi is also facilitated as the wealth of knowledge on the interface between glass and biological matter can be explored.

The electrical properties of PSi also differ strongly from those of bulk silicon [17]. Figure 3.5 shows results from time-of-flight measurements performed on oxidized PSi [18]. These results suggest that transport in this type of PSi is more similar to transport in a-Si:H than in crystalline Si. The results are consistent with a model that invokes transport on a fractal. Other transport mechanisms

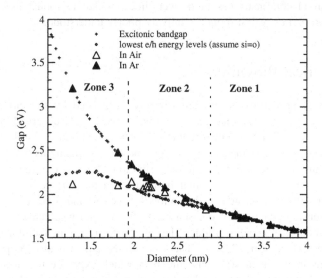

FIGURE 3.4. Calculated excitonic band gap for silicon quantum dots of different sizes. When one Si-H passivating bond is replaced by a Si=O bond, the gap does not increase as fast when the diameter decreases. The data points show photoluminescence measurements performed in argon and in air.

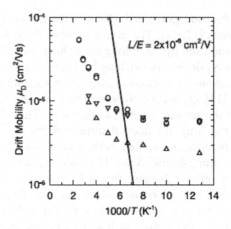

FIGURE 3.5. Drift mobility measured on an oxidized PSi pin diode structure as a function of temperature for a given ratio L/E of the carrier displacement over the applied electric field. The circles correspond to electron transport and the triangles to hole transport. The solid line shows data obtained for holes in a-Si:H.

may be in play as the connectivity of the porous structure and the average size of the silicon remnants change [17]. As we will discuss in the next section, the electrical properties of PSi, such as the conductivity, are also strongly influenced by environmental conditions, including adsorption of molecules or the presence of a fluid inside the pores. In contrast to the situation with optical properties, where a broad consensus has built over time and led to robust models, the electrical properties of PSi remain relatively poorly understood.

3.2 Electrical Biosensors

The electrical properties of PSi are strongly influenced by the environment as evidenced by the large changes in conductivity [19–21] observed when the porous layer is infiltrated with organic solvents or exposed to molecules. At least two response mechanisms have been proposed to explain this behavior. The first mechanism suggests that charge redistribution takes place in the crystallites as a result of the alignment of polar molecules on the surface [22, 23]. This mechanism will be described in the following section and used to explain the electrical sensing principle of our sensor. In the second mechanism, oxidation and adsorption of molecules produce charge transfer reactions facilitated by the presence of surface traps [23, 24]. The second mechanism is also supported by observations made on mesoporous silicon in which exposure to valence band and conduction band quenchers with different redox potentials produces a shift in the photoluminescence (PL) peak [25].

Changes in capacitance and conductance of PSi have been widely used as transduction mechanisms in various sensing situations [22,23,26–33], but limited work has been done in the field of biological electrical sensors. Different electrode configurations have been reported to measure the changes in electrical properties of PSi. In the most common situation, metal contacts are evaporated on the PSi surface [34] with various configurations such as interdigitated and coplanar electrodes [24, 35, 36] or in a top and bottom ("sandwich") configuration [22, 23, 33]. If the electrodes are in contact with the solvent or the buffer, the device characteristics may change over time as degradation at the contact electrodes is likely. Therefore, the electrodes need to be protected. An alternative geometry consists of using backside coplanar electrodes placed on the c-Si substrate and rely on the propagation of the electric field from the c-Si substrate to the PSi layer [37]. In this case the influence of the surface/metal contact barrier and any chemical reaction between the sensing species and the surface of the porous layer is eliminated. The response of these sensors is therefore related to the interaction of the sensing species and the PSi surface and can be explained with a Space Charge Region Modulation (SCRM) model.

3.2.1 Sensing Principle

Figure 3.6 shows a schematic representation of the PSi electrical sensor with backside contacts connected to a computer-controlled LCR meter. When the porous layer is infiltrated with different liquids, a change in the baseline capacitance (C) and conductance (G) can be measured in real time. The sensor can be modeled as a field effect transistor (FET) in which the metal gate is substituted by the porous layer that acts as a chemical gate. The sensing principle is explained on the basis of the SCRM model [37]. When the surface of the porous layer is exposed to a solution, the liquid/solid interface tends to reach electrostatic equilibrium by modifying the space charge region in the PSi walls. An electrical double layer is formed between the space charge region (silicon) and the Helmholtz layer, which is sensitive to the adsorption/desorption of ions and the components of the liquid. Changes in the Helmholtz layer lead to a

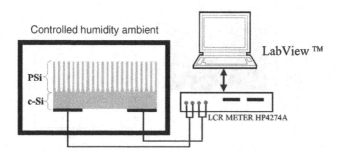

FIGURE 3.6. Schematic of a PSi electrical sensor with the backside contacts connected to a computer-controlled LCR meter.

modification of the space charge region in the silicon walls as a new electrostatic equilibrium is reached.

According to the model, adsorbed molecules will change the space charge region of the PSi walls and the underlying c-Si substrate. Since the doping of the substrate is relatively low, a small change of the surface charge has a strong influence on the depletion region width. The degree to which this change occurs will depend on the physical properties of the adsorbed molecule (e.g., dipole moment, polarizability). After infiltration with a liquid, the porous layer behaves as a charged layer that modifies the field distribution in the whole structure (PSi/c-Si). This change in the electric field distribution of the structure is measured in real time as a change in capacitance (C) and conductance (G) and the device can be modeled with the simplified equivalent electrical circuit shown in Figure 3.7.

The porous layer is represented by two RC components, one for the pores, which changes with the dielectric constant of the liquid, and another for the pore walls, which is related to changes in the depletion layer induced by the adsorption of molecules. The c-Si substrate is represented by a variable conductance that changes depending on the charge redistribution in the walls of the porous layer. From the point of view of an FET, the variable c-Si conductance represents the channel conductance.

3.2.2 Flow-Through Sensor Fabrication

A flow-through biosensor would be more desirable than the biosensor configuration shown in Figure 3.6, as larger volumes of liquid can be passed through the sensor, leading to increased response speed and enhanced sensitivity [38, 39].

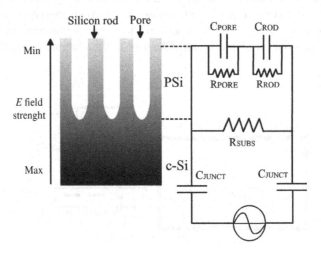

FIGURE 3.7. Schematic representation of the cross section of the PSi sensor and the electrical equivalent circuit with the associated capacitive and conductive components.

A schematic representation of a flow-through device structure is shown in Figure 3.8. This sensor configuration isolates the contacts, and, potentially, microelectronics circuitry from the backside fluidics. Note that the contact regions are not within the flow-through regions of the membrane. A protecting layer (i.e., SiO_2, polyimide) can be easily patterned over the metal electrodes to avoid topside interaction with fluids. The lateral extension of the PSi beyond the contact regions minimizes the shunt conductance through the silicon substrate. The V-groove produced by KOH etching can extend over an array of devices, but the lateral etching of the PSi membrane beyond the mask edge presents scaling limitations.

Macroporous silicon electrical sensors were fabricated using an electrolyte with 4 wt% hydrofluoric acid (49 wt%) in N, N-dimethylformamide (DMF). The use of a mild oxidizer such as DMF with low-doped silicon results in very straight pore walls and pore diameters between 1 and 2 μm [40]. The membranes were etched to a depth of approximately 70 μm using silicon nitride as an etch mask. Another nitride mask was defined on the backside of the wafer (aligned to the front) and a KOH etch was used to form a deep cavity, providing fluidic delivery to the free-standing membrane. With this procedure, an overetch can be used to compensate for pore depth nonuniformity due to current-crowding effects during the electrochemical etching process. Finally, the structure was thermally oxidized, providing a SiO_2 layer approximately 1000 Å thick within

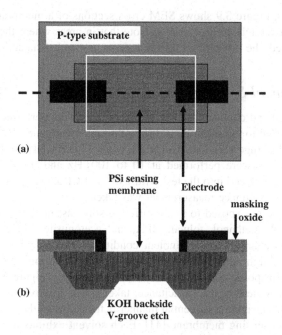

FIGURE 3.8. Layout (a) and cross section (b) of an integrated PSi-based flow-through sensor device. The dashed line across (a) shows the cut –line for the cross section in (b).

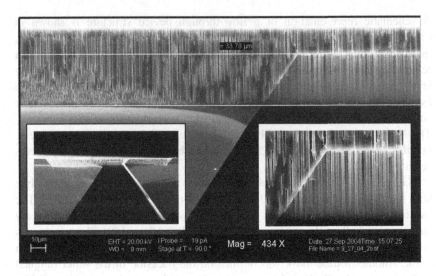

FIGURE 3.9. SEM cross sections of the macroporous silicon flow-through sensing membrane. The background image is a magnification of the inset on the left. The PSi layer was etched to a depth $\sim 50\,\mu$m. The KOH backside etch continued beyond the intersection with the porous layer. Pore liners of silicon dioxide and silicon nitride remain suspended beneath the remaining 20-μm-thick PSi membrane.

the macropores. Figure 3.9 shows SEM cross sections of a membrane structure that has been overetched considerably. Note that in areas where the membrane has been thinned, the protecting liners remain as suspended stalactites.

3.2.3 Chemical Sensing

Capacitance (C) and conductance (G) measurements were performed with both PSi layers attached to the substrate and flow-through membranes. Real-time data acquisition and storage was done using a PC control interfaced with LabView™. The measurements were performed at 10 to 100 kHz and 0 V DC bias. The device chips were placed into the test fixture of the LCR meter, channel side up, and a baseline capacitance measurement was taken.

The sensors were exposed to many different solvents, including chloroform, acetone, ethanol, methanol, toluene, IPA, and acetonitrile. Since the sensors are sensitive to changes in the ambient conditions such as relative humidity, temperature, and light, these parameters were monitored during testing to ensure that the device response was exclusively from the analyte. Figure 3.10 shows a real-time measurement of the capacitance, taken when approximately 30 μL of solvent was introduced in a sequential manner in the center of the channel, filling the channel and sensing membrane [41]. Each solvent exhibits a characteristic signature (amplitude, sign, shape, duration). Each of these features is related to the different physical properties of the solvent such as vapor pressure, dipole

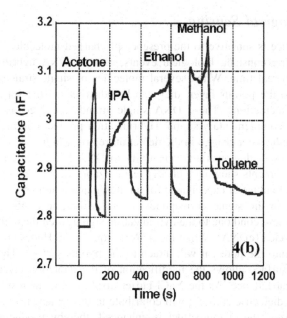

FIGURE 3.10. Characteristic signatures of five solvents tested sequentially in a flow-through sensor. Each signature is unique to the specific solvent, with distinctive features (shape, duration). Although there is a baseline drift, the response signatures are virtually identical to those taken with sensors dedicated to a single solvent.

moment, dielectric constant, and polarizability. Water and nonpolar solvents exhibit a different response which can be used to identify each of them [37]. After exposure to toluene, the sensor permanently loses its sensitivity. A demonstration of the repeatability and reproducibility of the sensors is shown in Figure 3.11.

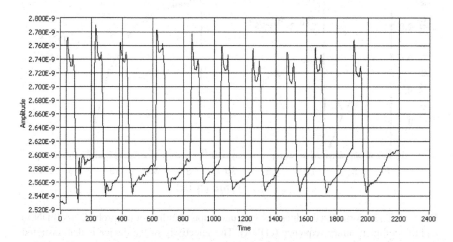

FIGURE 3.11. Successive tests with methanol on the same sensor.

3.2.4 Biological Sensing

Since the device is sensitive to the presence of charged molecules and changes in the dielectric constant, biological events, such as DNA hybridization, can be detected in real time. When several molecules of single-stranded DNA are infiltrated into the porous layer, they can be adsorbed on the surface through electrostatic interactions. These DNA single strands are used as probes that can capture a complementary strand. Hybridization produces an increase in the overall charge hence a change in the depletion region width.

The PSi sensors demonstrated sensitivity to DNA hybridization, with capacitance measurements showing a clearly recognizable signal. Figure 3.12 shows the normalized values of capacitance for a binding and a nonbinding event [38]. The first rise in the signal corresponds to the infiltration of the probe DNA (pDNA) and the second one to the hybridization signal when the complementary sequence is added (cDNA). When the pDNA sequence in Hepes buffer or NaCl is infiltrated into the pores, it will attach to the porous surface. The thin oxide layer on the surface allows the modification of the depletion layer leading to a change in capacitance. As the NaCl buffer displaces the air inside the pores, the change in dielectric constant will contribute to the capacitance signal. When DNA hybridizes, the charge effect is enhanced, thereby producing a rise in the signal above the one produced by the probe DNA (pDNA). The principle of detection of the device is based on the simultaneous effect of a change in dielectric constant and a modulation of the depletion layer of the structure resulting from the presence of charged molecules, in agreement with the model of Section 3.2.1. To assess the selectivity of the device we performed the same

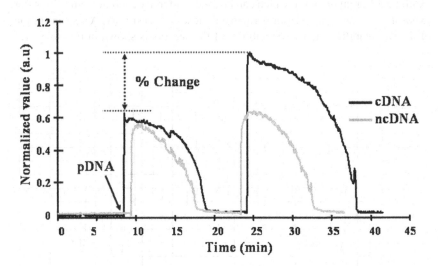

FIGURE 3.12. Normalized values of capacitance upon binding of the probe DNA (pDNA) and the complementary sequence (cDNA). The selectivity of the device is demonstrated by exposing pDNA to a noncomplementary sequence (ncDNA).

experiment with a noncomplementary sequence. In this case, although a signal is produced, the change in C when the noncomplementary DNA (ncDNA) is infiltrated is significantly less ($\sim 10\%$) than the one induced when binding occurs ($\sim 40\%$).

The effect of the charge of DNA alone on the response of the PSi sensors can be examined by using an uncharged analog of DNA known as peptide nucleic acid (PNA). Since PNA is composed of the same bases as DNA, it hybridizes to DNA but hybridization should not produce the same charge effect. Experimental results support this prediction. When PNA and DNA hybridize, the measured signal was found to be identical to that measured when a noncomplementary DNA strand is introduced into the pores [42]. Mass and charge effects on the response of the sensor were also studied by performing various control experiments. For instance, while the infiltration of the buffer itself produces a change in impedance and phase angle, the magnitude is significantly different from that of the binding event. In this specific case the response is also produced by the adsorption of the buffer components on the surface as well as a change in the dielectric constant of the pores.

3.3 Optical Biosensors

PSi is a good host material for label-free optical biosensing applications because its optical properties (photoluminescence, reflectance) are highly sensitive to the presence of chemical and biological species inside the pores [43]. PSi-based affinity optical biosensors with a variety of configurations such as single layer Fabry-Perot cavities [44], Bragg mirrors [45], rugate filters [46], and microcavities [47–49] have been experimentally demonstrated for the detection of toxins [50], DNA [47], and proteins [10, 44]. The capture of biological or chemical molecules inside the pores increases the effective refractive index of the PSi structures thus causing a red shift in the photoluminescence or reflectance spectra.

3.3.1 Sensing Principle

The effective dielectric constant of a PSi layer is related to its porosity by the Bruggeman effective medium model [11]:

$$(1-P)\frac{\varepsilon_{Si}-\varepsilon_{PSi}}{\varepsilon_{Si}+2\varepsilon_{PSi}}+P\frac{\varepsilon_{void}-\varepsilon_{PSi}}{\varepsilon_{void}+2\varepsilon_{PSi}}=0 \qquad (3.1)$$

where P is porosity, ε_{PSi} is the effective dielectric constant of porous silicon, ε_{Si} is the dielectric constant of silicon, and ε_{void} is the dielectric constant of the medium inside the pores. The Bruggeman effective medium model shows that the effective refractive index of PSi ($n_{eff}^2 = \varepsilon_{PSi}$) increases as the porosity decreases and as $\varepsilon_{\text{void}}$ increases. In Figure 3.13, n_{eff} is plotted as a function of porosity [10]. In sensing applications, $\varepsilon_{\text{void}}$ increases due to the binding of targets to the internal surface of the pores. Thus, the overall effective dielectric constant of the porous structure ε_{PSi} increases which causes a red shift in the optical reflectance spectrum of the structure. As shown in Figure 3.13, for a given increase of $\varepsilon_{\text{void}}$, the effective refractive index change is larger for higher porosity layers.

In order to selectively detect targets of interest, the internal surface of PSi needs to be functionalized. Highly selective elements, such as DNA segments and antibodies, can be immobilized on the internal surface of the pores as the bioreceptors or probe molecules. When the sensors are exposed to the target, the probe molecules selectively capture the target molecules. The molecular recognition events are then converted into optical signals via the increase of the refractive index.

A PSi microcavity is a one-dimensional photonic band gap structure that contains a defect (symmetry breaking) layer sandwiched between two Bragg mirrors [51]. Each Bragg mirror is a periodic stack of layers with two different porosities and quarter wavelength optical thickness. The schematic drawing of a PSi microcavity and its typical reflectance and photoluminescence spectra are shown in Figure 3.14. Depending on the thickness of the defect layer, the reflectance spectrum of a microcavity is characterized by one or several sharp resonance dips in the stop band. The reflectance spectrum modulates the measured PL spectrum and produces one or several very narrow PL peaks. The position of the reflectance dip or photoluminescence peak is determined by the optical thickness (refractive index times the physical thickness) of each layer in the structure. A slight change of the refractive index inside the pores causes a shift of the spectrum.

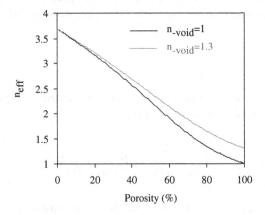

FIGURE 3.13. Effective refractive index of PSi as a function of the porosity at the wavelength of 800 nm. The black curve corresponds to $n_{\text{void}} = 1$ and the gray curve to $n_{\text{void}} = 1.3$.

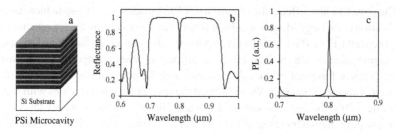

FIGURE 3.14. (a) Schematic of a PSi microcavity. (b) Simulation of the reflectance spectrum of a PSi microcavity with a half-wavelength optical thickness defect layer and 6-period Bragg mirrors. (c) Simulation of the PL spectrum of the same microcavity.

3.3.2 Microcavity Sensor Design and Sensitivity

The figure of merit describing the sensitivity of affinity sensors is $\Delta\lambda/\Delta n$, where $\Delta\lambda$ is the wavelength shift and Δn is the change of the ambient refractive index. For a PSi microcavity sensor with an average porosity of 75%, $\Delta\lambda/\Delta n$ is ~ 550 nm [52], which is much larger than for sensing platforms that rely on the interaction between the evanescent tail of the field and the analyte [53]. For a system able to detect a wavelength shift of 0.1 nm, the minimum detectable refractive index change of the pores is 2×10^{-4}.

The sensitivity of PSi microcavity sensor can be increased by increasing its Q-factor ($Q = \lambda/\Delta\lambda$, where λ is the resonance center wavelength and $\Delta\lambda$ is the full width half maximum of the resonance dip). For higher Q-values, the spectral features are sharper and smaller shifts can be detected. The Q-factor can be increased by increasing the contrast in the porosity of the layers. However, a higher contrast in porosity is usually produced by a higher contrast in pore opening, which may not be favorable for biosensing applications when easy infiltration of the target throughout the entire multilayer structure is required. For a given porosity contrast, the Q-factor can also be increased by increasing the number of periods of the Bragg mirror. In practice, the number of periods cannot be increased arbitrarily for two reasons. First, uniform infiltration of the molecules becomes more difficult in thicker devices. Second, maintaining a constant HF concentration at the tip of very deep pores is difficult [54], which may lead to an undesired porosity or refractive index gradient.

PSi-based PBG sensors also have the advantage of size exclusion. When the PSi sensor is exposed to a complex biological mixture, only the molecules that are smaller than the pores can be infiltrated into the sensor. Furthermore, an increase of refractive index on the top of the microcavity due to the nonspecific binding of large size unwanted objects only causes changes to the side lobes in the reflectivity spectrum, not to the resonance dip [55]. Thus, PSi microcavities are more reliable than planar sensing platforms, where the nonspecific binding of large-size objects present in a "dirty" environment may produce a false-positive signal.

The pore size also affects the sensitivity of PSi biosensors because in biosensing applications, the target does not completely fill the pores but instead is attached to the pore walls. For a PSi layer of a given porosity, the internal surface area decreases as the pore size increases. The effective refractive index change of a layer with larger pores is thus smaller since the percentage of the pore volume occupied by the biological species is smaller. We simulated the spectra for microcavities with fixed porosities (80% for the high-porosity layer and 70% for the low-porosity layer) but different pore diameters ranging from 20 nm to 180 nm. For a given thickness of the coating (with $n_{layer} = 1.42$, a typical value for biomolecules), the resonance red shift decreases as the pore size increases, as shown in Figure 3.15 [10]. It can be seen that, for a 1-nm-thick coating layer, a microcavity with 40-nm pores produces a red shift of 23 nm, while a microcavity with 100-nm pores gives a red shift of only 10 nm. Thus, to optimize the sensitivity, the pore size should be as small as possible while still allowing for easy infiltration of the biological material.

From Figure 3.15 we can also estimate the areal mass sensitivity in terms of gram/internal surface area. We assume a detection system able to resolve a shift of 0.1 nm. For a microcavity with 20-nm pores, the minimum coating thickness required to achieve a detectable signal is 0.002 nm, which is equivalent to $\sim 2\,pg/mm^2$. For a microcavity with 180-nm pores, the minimum coating thickness is 0.02 nm, which is equivalent to $\sim 20\,pg/mm^2$.

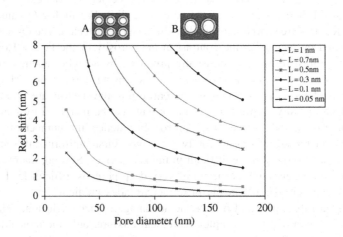

FIGURE 3.15. Simulation of microcavity spectrum red shift as a function of pore size. Each curve represents a different thickness (L) of the coating layer with $n = 1.42$. The porosities of the layers used in all the calculations were fixed at 80% and 70%. For a given thickness of the coating (shaded area), the effective refractive index change of a layer with small pores (A) is higher than that of a layer with large pores (B), which causes a larger shift in the spectrum. For a given pore size, the thicker the coating, the larger the shift.

3.3.3 Fabrication of PSi Microcavity

The porosity can be controlled by the etching current density. Once a layer has been etched, further etching using a different current density does not affect it as explained in Section 3.1.1. Stacks of layers with different refractive indices can thus be formed by alternating the current density during etching [56]. Figure 3.16 shows a cross-sectional SEM image of a multilayer structure etched by a periodic current density pulse train. Each specific current density produces a different porosity. The duration of current density determines the thickness of the layer.

By controlling the current density and etching time, layers with the proper optical thickness can be formed. Figure 3.17 shows cross-sectional and top-view SEM images of two different types of microcavities with different pore sizes. The mesoporous silicon microcavity with pore diameters from 10 nm to 50 nm was fabricated from a p+ wafer (0.01 ohm-cm) with 15% HF in ethanol. The macroporous silicon microcavity with pore diameters from 80 nm to 150 nm was etched in an n+ wafer (0.01 ohm-cm) with 5.5% HF in DI water [10]. The mesoporous microcavity is a good option for detection of small-size targets such as short DNA strands, while the macroporous microcavity is suitable for macromolecule sensing.

3.3.4 Biological Sensing

3.3.4.1 DNA Detection

Mesoporous silicon microcavities were investigated for the detection of small DNA segments (22 based pairs) [47, 57]. Strong photoluminescence is observed from the mesoporous silicon microcavity. Multiple photoluminescence peaks as narrow as 3 nm of full width half maximum (FWHM) were measured from a microcavity with thick defect layer. The first step in the sensor preparation

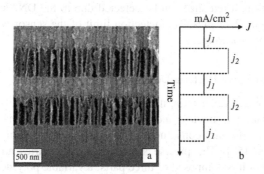

FIGURE 3.16. (a) Cross-sectional SEM image of a PSi multilayer structure formed by a periodic current pulse train. The porosity and thickness of each layer are determined by the current density amplitude and duration, respectively. (b) Schematic drawing of the etching current density over time.

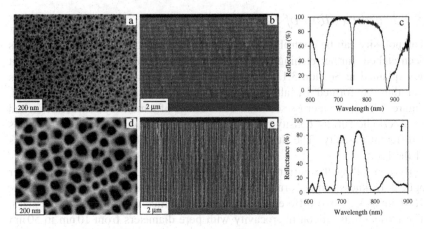

FIGURE 3.17. (a, b) Top-view and cross-sectional SEM images of a mesoporous silicon microcavity with pore sizes from 10 nm to 50 nm. (c) Reflectance spectrum of a similar mesoporous silicon microcavity. (d, e) Top-view and cross-sectional SEM images of a macroporous silicon microcavity with pore size from 80 nm to 150 nm. (f) Reflectance spectrum of the same macroporous silicon microcavity.

involves the silanization of the thermally oxidized porous silicon sample with 3-glycidoxypropyltrimethoxy silane, which is hydrolyzed to a reactive silanol by using DI water (pH ~4). The thermal oxidation treatment not only produces a silica-like surface, but also improves the photo-stability of luminescent porous silicon film (see Section 3.1.2). After successful silanization, the probe DNA, which has an amine group attached to the 3 end of the sequence, is immobilized onto the pore surface via diffusion. The amine group attacks the epoxy ring of the silane, thereby opening it up and forming a bond. Finally, the DNA attached porous silicon sensor is exposed to the complementary strand of DNA (cDNA). Comparing the photoluminescence spectrum of the sensor before and after the DNA hybridization, a red shift can be detected due to the DNA/cDNA binding as shown in Figure 3.18 [47]. The detection limit of the sensor was determined to be in the picomolar range.

3.3.4.2 Gram-Negative Bacteria Detection

To detect gram-negative bacteria, it is first necessary to select a target molecule present in this bacterial subclass and not in gram-positive bacteria. Lipopolysaccharide (LPS) is a primary constituent of the outer cellular membrane of gram-negative bacteria [58]. The precise structure of LPS varies among bacterial species but overall it is composed of three parts: a variable polysaccharide chain, a core sugar, and lipid A. As lipid A is common to all LPS subtypes, this seems a natural target. An organic receptor, tetratryptophan *ter*-cyclopentane (TWTCP), was designed and synthesized as the probe molecule. TWTCP specifically binds to diphosphoryl lipid A in water with a dissociation constant of 592 nm [59].

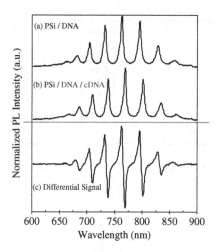

FIGURE 3.18. (a) PL spectrum of a mesoporous silicon microcavity with probe DNA (50 μM) immobilized inside the device. (b) PL spectrum of the microcavity after exposure to cDNA (1 μM). (c) Differential PL spectrum.

After the mesoporous silicon microcavity sensor is treated with 3-glycidoxy-propyltrimethoxy silane, a mixture of TWTCP and glycine methyl ester is applied to the sensor. Glycine methyl ester is used as blocker molecules to avoid the reaction of the four amino groups of the tetratryptophan receptor with the epoxide-terminated PSi surface. As shown in Figure 3.19, upon exposure of lysed gram-negative cells (*Escherichia coli*) to the immobilized TWTCP sensor, a 4-nm red shift of the photoluminescence spectrum of the microcavity is detected [60]. However, when the sensor is exposed to a solution of lysed gram-positive cells (*Bacillus subtilis*), no shift of the spectrum is observed. These results were confirmed with all gram-negative and gram-positive bacteria tested.

3.3.4.3 Protein Sensing

Mesoporous silicon microcavities are better suited for the detection of small objects. For large size protein detection, macroporous silicon microcavities are a better option because they allow much easier infiltration of large molecules such as immunoglobulin G (IgG). A protein biosensor based on a macroporous silicon microcavity was demonstrated with the streptavidin–biotin couple [10]. While biotin is a small molecule, streptavidin is relatively large (67 kDa), making its infiltration difficult into mesopores but easy into macropores. Furthermore, each streptavidin tetramer has four equivalent sites for biotin (two on each side of the complex), which makes it a useful molecular linker.

To create a biotin-functionalized sensor for the capture of streptavidin, the microcavity was first thermally oxidized, then silanized with aminopropyltri-ethoxysilane to create amino groups on the internal surface. The probe molecules,

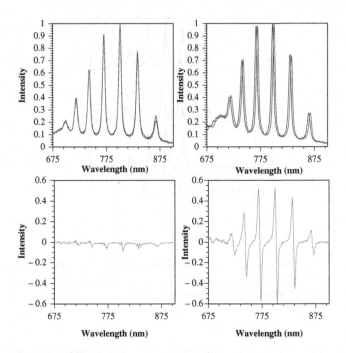

FIGURE 3.19. Photoluminescence spectra of a mesoporous silicon microcavity biosensor in the presence and absence of bacterial cell lysates.(top) The spectra on the left show response to gram-positive bacteria, while the spectra on the right show response to gram-negative bacteria.(bottom) The traces at the bottom are the differential spectra.

sulfo-NHS-LC-LC-biotin   [sulfosuccinimidyl-6-(biotinamido)-6-hexa-namido hexanoate], were then immobilized inside the pores. As shown in Figure 3.20a, the red shift of the resonance increases as the biotin concentration increases. Using the simulation discussed in Section 3.3.2, we can estimate the biotin surface coverage, which is linearly related to the red shift in Figure 3.20a. A 10-nm red shift corresponds to a nearly complete (95%) biotin surface coverage.

To study how the biotin surface coverage affects the binding to streptavidin, sensors derivatized with different biotin concentrations were exposed to the same amount of the target: $50\,\mu l$ of streptavidin with a concentration of 1 mg/ml. After exposure to streptavidin, a red shift was detected which is attributed to the specific binding of streptavidin to the biotin-derivatized macroporous microcavity (Figure 3.20b). For the control samples that were silanized but did not contain biotin, no shift was detected after exposure to streptavidin, which indicates that nonspecific absorption of streptavidin inside the microcavity is nonexistent or negligible.

The red shift caused by streptavidin binding to biotin is a function of the biotin surface coverage. Figure 3.20c shows that there is an optimum biotin surface coverage that maximizes the red shift, therefore the capture of streptavidin. This behavior is in good agreement with observations by other researchers using

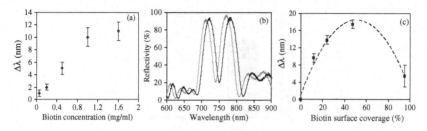

FIGURE 3.20. (a) Dependence of the red shift on the concentration of biotin applied on the sensor. (b) A 10-nm red shift is detected after the sensor is exposed to the target. The gray curve is the reflectance spectrum of the sensor with biotin. The black curve is the reflectance spectrum after the sensor is exposed to streptavidin (1 mg/ml). (c) Red shift due to specific binding of streptavidin as a function of the biotin surface coverage.

different sensing techniques [61]. The spacing between each neighboring biotin needs to be large enough so that biotin can reach the pocketlike binding site of streptavidin. The roll-off of the red shift shown in Figure 3.20c is due to the fact that when the biotin surface density becomes very high and each biotin is closely surrounded by its neighbors, biotin cannot protrude deep enough into the binding site of streptavidin, thus decreasing the probability of capturing streptavidin.

3.3.4.4 IgG Sensor

IgG is the most common type of antibody synthesized in response to a foreign substance (antigen). The antibody has a specific molecular structure capable of recognizing a complementary molecular structure on the antigen which might be some proteins, polysaccharides, and nucleic acids. From the X-ray crystal structure, the longest dimension of IgG is approximately 17 nm [62].

The detection of rabbit IgG (150 kDa) was investigated through multiple layers of biomolecular interactions in a macroporous silicon microcavity sensor. The silanized sensor was first derivatized with biotin which can selectively capture streptavidin as described in the previous section. The immobilized streptavidin can be used as a linker because of its free binding sites. Exposure of biotinylated goat anti-rabbit IgG to the sensor results in its attachment to the surface through the binding between biotin and streptavidin. The sensor uses goat anti-rabbit IgG as the probe molecule to selectively capture rabbit IgG. A red shift of the spectrum can be detected when each layer of molecules is added to the sensor. As shown in Figure 3.21, when the sensor was exposed to $50 \mu l$ of a solution containing rabbit IgG at a 1 mg/ml concentration, a 6-nm red shift was detected [52]. When the sensor was exposed to $50 \mu l$ goat IgG (1 mg/ml) which does not bind to the goat anti-rabbit IgG, the red shift is extremely small (< 0.5 nm). This control experiment shows that the sensor can selectively detect rabbit IgG.

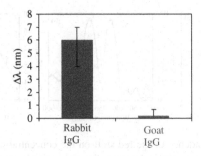

FIGURE 3.21. A 6-nm red shift of the spectrum is detected when the PSi microcavity sensor is exposed to the target molecule rabbit IgG. When the sensor is exposed to goat IgG, the red shift is extremely small.

3.4 Conclusions

Biological and chemical sensors made of porous silicon have been designed, fabricated, and tested. They rely on the sensitivity of the electrical and optical properties of this material to environmental factors. Models have been proposed to explain and quantify the response of the sensors to various chemicals and biological matter, including DNA and proteins. The fact that they are fabricated in silicon suggests that single sensor elements and sensor arrays can be fabricated inexpensively and reliably. We believe that these sensors are useful as analytical tools as well as for point of care applications and for use by untrained individuals.

Acknowledgments. This work was supported in part by grants from the U.S. National Science Foundation and the Infotonics Center of Excellence in Canandaigua, NY. The authors acknowledge the following individuals for their contributions: Professor B. Miller, Professor K. Hirschman, Dr. S. Chan, Dr. M. Christophersen, Dr. L. DeLouise, Dr. S. Horner, J. Clarkson, and R. Viard.

References

1. A. Uhlir, Jr., *Bell Syst. Tech. J.* **35**, 333 (1956).
2. Y. Watanabe, Y. Arita, T. Yokoyama, and Y. Igarashi, *J. Electrochem. Soc.* **122**, 1351 (1975).
3. P. M. Fauchet, in *Encyclopedia of Applied Physics*, Update 2, Wiley-VCH Verlag, pp. 249–272 (1999).
4. L. T. Canham, in *Properties of Porous Silicon*, edited by L. Canham, INSPEC, London, pp. 81–89 (1997).
5. L. T. Canham, *Appl. Phys. Lett.* **57**, 1046 (1990).
6. V. Lehmann and U. Gosele, *Appl. Phys. Lett.* **58**, 856 (1991).
7. R. L. Smith and S. D. Collins, *J. Appl. Phys.* **71**, R1 (1992).
8. A. G. Cullis, L. T. Canham, and P. D. J. Calcott, *J. Appl. Phys.* **82**, 909 (1997).
9. V. Lehmann, *Electrochemistry of Silicon*, Wiley-VCH Verlag, GmbH (2002).

10. H. Ouyang, M. Christophersen, R. Viard, B. L. Miller, and P. M. Fauchet, *Adv. Funct. Mater.* **15**, 1851 (2005).

11. W. Theiss, *Surf. Sci. Rep.* **29**, 91 (1997).

12. P. M. Fauchet, in *Light Emission in Silicon: From Physics to Devices*, edited by D. J. Lockwood, Academic Press, San Diego, pp. 206–252 (1998).

13. M. S. Hybertsen, *Phys. Rev. Lett.* **72**, 1514 (1994).

14. M. V. Wolkin, J. Jorné, P. M. Fauchet, G. Allan, and C. Delerue, *Phys. Rev. Lett.* **82**, 197 (1999).

15. R. T. Collins, M. A. Tischler, and J. H. Stathis, *Appl. Phys. Lett.* **61**, 1649 (1992).

16. L. Tsybeskov, S. P. Duttagupta, K. D. Hirschman, and P. M. Fauchet, *Appl. Phys. Lett.* **68**, 2058 (1996).

17. A. J. Simons, in *Properties of Porous Silicon*, edited by L. Canham, INSPEC, London, pp. 163–184 (1997).

18. P. N. Rao, E. A. Schiff, L. Tsybeskov, and P. Fauchet, *Chem. Phys.* **284**, 129 (2002).

19. C. Cadet, D. Deresmes, D. Villaume, and D. Stievenard, *Appl. Phys. Lett.* **64**, 2827 (1994).

20. V. Lehmann, F. Hofmann, F. Moeller, and U. Gruning, *Thin Solid Films* **255**, 20 (1995).

21. L. A. Balagurov, D. G. Yarkin, and E. A. Petrova, *Mater. Sci. Eng. B* **69/70**, 127 (2000).

22. M. Ben-Chorin, A. Kux, and I. Schechter, *Appl. Phys. Lett.* **64**, 481 (1994).

23. I. Schechter, M. Ben-Chorin, and A. Kux, *Anal. Chem.* **67**, 3727 (1995).

24. S. Green and P. Kathirgamanathan, *Mater. Lett.* **52**, 106 (2002).

25. J. Rehm, G.L. McLendon, and P.M. Fauchet, *J. Am. Chem. Soc.* **118**, 4490 (1996).

26. G. Barillaro, A. Nannini, and F. Pieri, *Sens. Actuat. B* **93**, 263 (2003).

27. S.-J. Kim, S.-H. Lee, and C.-J. Lee, *J. Phys. D* **34**, 3505 (2001).

28. M. Thust, M. J. Schöning, S. Frohnoff, R. Arens-Fischer, P. Kordos, and H. Lüth, *Meas. Sci. Technol.* **7**, 26 (1996).

29. M. Ben Ali, R.Mlika, H. Ben Ouada, R. M'ghaïeth, and H. Maâaref, *Sens. Actuat. A* **74**, 123 (1999).

30. R. R. K. Reddy, A. Chadha, and E. Bhattacharya, *Biosens. Bioelectron.* **16**, 313 (2001).

31. Z. M. Rittersma, W. J. Zaagman, M. Zetstra, and W. Benecke, *Smart Mater. Struct.* **9**, 351 (2000).

32. S.-J. Kim, J.-Y. Park, S.-H. Lee, and S.-H. Yi, *J. Phys. D* **33**, 1781 (2000).

33. A. Motohashi, M. Ruike, M. Kawakami, H. Aoyagi, A. Kinoshita, and A. Satou, *Jpn. J. Appl. Phys.* **35**, 394 (1996).

34. R. C. Anderson, R. S. Muller, and C. W. Tobias, *Sens. Actuat. A* **A21/A23**, 835 (1990).

35. K. Watanabe, T. Okada, I. Choe, and Y. Sato, *Sens. Actuat. B* **33**, 194 (1996).

36. S.-J. Kim, S.-H. Lee, and C.-J. Lee, *J. Phys. D* **34**, 3505 (2002).

37. M. Archer, M. Christophersen, and P.M. Fauchet, *Sens. Actuat. B* **106**, 347 (2005).

38. M. Archer, M. Christophersen, P. M. Fauchet, D. Persaud, and K. D. Hirschman, *Mater. Res. Soc. Symp. Proc.* **782**, 385 (2004).

39. J. P. Clarkson, V. Rajalingam, K. D. Hirshman, H. Ouyang, W. Sun, and P. M. Fauchet, *Mater. Res. Soc. Symp. Proc.* **869**, D321 (2005).

40. H. Foell, M. Christophersen, J. Carstensen, and G. Hasse, *Mater. Sci. Eng.* **R39**, 93 (2002).

41. J. P. Clarkson, P. M. Fauchet, V. Rajalingam, and K. D. Hirshman, IEEE Sensors Journal (to appear in March 2007).

42. M. Archer, M. Christophersen, and P.M. Fauchet, *Biomed. Microdev.* **6**, 203 (2004).

43. M. P. Stewart and J. M. Buriak, *Adv. Mater.* **12**, 859 (2000).

44. K.-P. S. Dancil, D. P. Greiner, and M. J. Sailor, *J. Am. Chem. Soc.* **121**, 7925 (1999).

45. P. A. Snow, E. K. Squire, P. S. J. Russell, and L. T. Canham, *J. Appl. Phys.* **86**, 1781 (1999).

46. F. Cunin, T. A. Schmedake, J. R. Link, Y.Y. Li, J. Koh, S. N. Bhatia, and M. J. Sailor, *Nature Mater.* **1**, 39 (2002).

47. S. Chan, Y.Li, L. J. Rothberg, B. L. Miller, and P. M. Fauchet, *Mater. Sci. Eng. C* **15**, 277 (2001).

48. L. Pavesi, *Riv. Nuovo Cimento* **20**, 1 (1997).

49. H. Ouyang, M. Christophersen, and P. M. Fauchet, *Phys. Status Solidi A* **202**, 1396 (2005).

50. H. Sohn, S. Letant, M. J. Sailor, and W. C. Trogler, *J. Am. Chem. Soc.* **122**, 5399 (2000).

51. J. E. Lugo, H. A. Lopez, S. Chan, and P. M. Fauchet, *J. Appl. Phys.* **91**, 4966 (2002).

52. H. Ouyang and P. M. Fauchet, *Proc. SPIE* **6005**, 600508-1 (2005).

53. J. Scheuer, W. M. G. Green, G. A. DeRose, and A.Yariv, *IEEE J. Sel. Top. Quantum Electron.* **11**, 476 (2005).

54. M. Thonissen, M. G. Berger, S. Billat, R. Ares-Fischer, M. Kruger, H. Luth, W. Theiss, S. Hillbrich, P. Grosse, G. Lerondel, and U. Frotscher, *Thin Solid Films* **297**, 92 (1997).

55. H. Ouyang, L. A. DeLouise, M. Christophersen, B. L. Miller, and P. M. Fauchet, *Proc. SPIE* **5511**, 71 (2004).

56. L. Pavesi, *Riv. Nuovo Cimento* **20**, 1 (1997).

57. S. Chan, P. M. Fauchet, Y. Li, L. J. Rothberg, and B. L. Miller, *Phys. Status Solidi A* **182**, 541 (2000).

58. L. S. Young, W. J. Martin, R. D. Meyer, R. J. Weinstein, and E. T. Anderson, *Ann. Intern. Med.* **86**, 456 (1997).

59. R. D. Hubbard, S. R. Horner, and B. L. Miller, *J. Am. Chem. Soc.* **123**, 5811 (2001).

60. S. Chan, S. R. Horner, P. M. Fauchet, and B. L. Miller, *J. Am. Chem. Soc.* **123**, 11797 (2001).

61. L. S. Jung, K. E. Nelson, P.S. Stayton, and C.T. Campbell, *Langmuir* **16**, 94221 (2000).

62. E. O. Saphire, P. W. H. I. Parren, R. Pantophlet, M. B. Zwick, G. M. Morris, P. M. Rudd, R. A. Dwek, R. L. Stanfield, D. R. Burton, and I. A. Wilson, *Science* **293**, 1155 (2001).

4

Optoelectronic-VLSI: Device Design, Fabrication, and Performance

David V. Plant

Department of Electrical and Computer Engineering, McGill University, Montreal, Quebec H3A 2A7

4.1 Introduction

This chapter describes design rules, discusses fabrication steps, and highlights the performance of a number of successful Optoelectronic-VLSI devices that have recently been demonstrated. Motivation for this work can be found by examining the Semiconductor Industry Association (SIA) roadmap which suggests that electrical interconnect technology for both on-chip and off-chip communications will be exhausted in the near future [28]. Optically interconnected chips, boards, and backplanes offer a potential solution to this anticipated electrical interconnect bottleneck. In this context there exist various approaches to quantifying the merits of optics versus electronics for distances ranging from millimeters to meters. A common design assumption is that optics provides increased connectivity through parallelism. Specifically, parallel optical interconnects are capable of providing high-bandwidth communication links both within and between high-performance electronic systems. The advantages of optical communications for long-distance interconnects are well known and provide the motivation for modern optical fiber networks. Optics is now challenging copper at shorter and shorter distances. The benefits of optical interconnects include reduced signal distortion and attenuation, lower power requirements, lighter components, potentially lower costs, and much greater immunity to electromagnetic interference. A thorough review of these physical issues is provided elsewhere [9, 10, 20, 22, 30]. In the commercial arena, several manufacturers now supply optical fiber ribbon-based parallel optical data links (PODLs) of 8 to 12 channels, operating at data rates of up to 3.25 Gb/s per channel over distances of 100–1000 m (depending on bit rate). However, there are applications where many more parallel channels are required and in some cases the interconnect configuration is more complex than a simple

73

point-to-point link. For example, multiprocessor computers, telecommunications switches, and embedded systems all require highly parallel interconnections.

In particular, the concept of the direct sourcing and termination of optical signals on silicon has been proposed as a method to relieve the off-chip communication bottleneck [36]. The objective of the approach is to create optical inputs and outputs (I/O) that operate in conjunction with traditional electrical I/O. In recent years, significant progress has been made in the enabling technologies required to achieve this objective. A pivotal step in this evolution includes successful demonstrations of the hybrid integration of Vertical Cavity Surface Emitting Lasers (VCSELs) and PIN photodiodes with Application Specific Integrated Circuits (ASICs) [13, 23, 32]. This emerging technology is now commonly referred to as Optoelectronic-VLSI (OE-VLSI). In this chapter, the author seeks to highlight the significant progress being made in Optoelectronic-VLSI technology with a focus on the enabling nanoscale optoelectronic devices and circuits. Much of the reported progress is as a result of a series of successful OE-VLSI ASIC demonstrations. Through the realization of these devices, research in a wide range of areas has been accomplished, including optoelectronic device design, electronic circuit design, and optical design and packaging.

This chapter is organized as follows. Section 4.2, OE-VLSI ASIC Design Space, is an overview of key design space criteria. Section 4.3, Heterogeneous Integration, is a description of the key materials parameters and fabrication requirements necessary for realizing OE-VLSI ASICs. Section 4.4, OE-VLSI ASIC Architectures, outlines the options for building these devices. Section 4.5, Enabling Analog Circuit Designs and Performance, summarizes suitable transceiver circuits that have been built using nanoscale CMOS technologies. Concluding remarks are given in Section 3.4.6.

4.2 OE-VLSI ASIC Design Space

In developing OE-VLSI technology, several assumptions were made: (1) the optical source is a VCSEL and the detectors are either PN or PIN devices. Most research in this area has been based on the use of devices operating at 850 and 960 nm, although many of the guidelines can be adapted to other wavelengths. (2) Heterogeneous integration using flip-chip bonding and substrate removal of optoelectronic devices onto foundry CMOS ASICS is used. (3) The optical interconnect distance is less than 150 mm (i.e., characteristic of interchip and board-to-board spacing). (4) The optical interconnects could be either fiber based or free-space based. (5) The interconnect is a point-to-point transmissive topology. (6) The interconnect incorporates a realistic tolerance to misalignment.

Identification of the design space in which OE-VLSI based chip-to-chip and board-to-board communications using optics was performed. Since the technology is aimed at future high-performance electronic systems, the International Technology Road Map for Semiconductors [28] was consulted. Figure 4.1

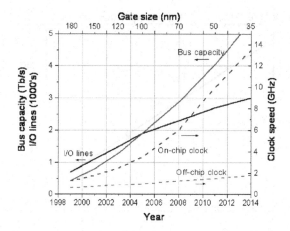

FIGURE 4.1. Projected evolution of on-chip clock speed (dashed line, right axis), off-chip clock speed (dotted line, right axis), number of high-speed I/O lines (dash-dot line, left axis), and total bus capacity (solid line, left axis) as a function of time and transistor size for high-performance systems.

shows the projected increase in VLSI transistor on-chip clock speed, off-chip clock speed, number of high-speed off-chip clock lines, and total off-chip I/O capacity as a function of time for high-performance systems, taken from reference [28]. It can be seen that by 2014 the off-chip clock speed is projected to reach 1.8 GHz and the width of the off-chip bus is also projected to increase to 3000 high-speed lines, with a total projected off-chip I/O capacity of 5 Tb/s. The on-chip clock speed is projected to reach 13.5 Gb/s. From these data, a view of the design space for off-chip interconnects by the year 2014 was obtained, under the assumption that the necessary off-chip bandwidth will be 5 Tb/s. This is shown in Figure 4.2, where the data rate per channel necessary to achieve 5 Tb/s is plotted as a function of the number of parallel channels.

From Figure 4.2, it is apparent that, assuming the optical interconnect channel data rate is not to exceed the projected on-chip data rate of 13.5 Gb/s (which would otherwise require the use of serialization/deserialization circuits), a 5 Tb/s aggregate data rate implies the presence of 370 optical I/Os. Assuming the off-chip optical links run at the projected electrical off-chip clock speed of 1.8 Gb/s implies approximately 2700 optical I/Os would be needed. Let these two values represent the boundaries for optical solutions to the off-chip interconnect problem. To this end, two recently reported OE-VLSI ASICs with 256 optical I/O [23] and 1080 optical I/O[32], respectively, address the required parallelism range. Other researchers have demonstrated matrix addressed VCSEL arrays with 4096 outputs [6].

To illustrate the utility of these OE-VLSI ASICs in an optical interconnect, an example of an experimentally realized point-to-point free-space optical interconnect is shown in Figure 4.3a (Châteauneuf 2002). In this system, two OE-VLSI ASICs were bidirectionally interconnected over a distance of

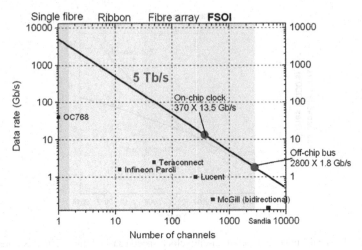

FIGURE 4.2. Projected off-chip I/O requirements for 2014.

3 inches. Each chip contains 256 transmitters (VCSELs) and 256 PINs. The interconnection distance was selected such that by inserting two additional prisms into the beam path, it could interconnect two boards in a bookshelf configuration. The relay optics was based on clustered diffractive lenses and the array density of 28 channels/mm^2 was achieved. This system represents the highest density free-space interconnect reported to date. A photograph of this system is shown in Figure 4.3b. This system illustrates the necessary components of an optical interconnect, including the ASICs to be interconnected, a PCB (or MCM substrate) on which the ASICs reside, the optical system used for interconnection, and the optical packaging required to realize the link.

In designing such a system, some of the OE-VLSI ASIC issues which must be addressed are: materials and integration, CMOS compatibility, transceiver architectures, digital functionality, relay optics, and the various packaging levels, including electrical packaging (which concerns issues such as signal integrity and power dissipation), OE-VLSI ASIC packaging, and optical packaging (which

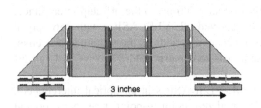

FIGURE 4.3. (a) Representative free-space optical interconnect. (b) Photograph of an experimentally realized 512-channel interchip interconnect system.

concerns the assembly of the relay optics). Finally, it is necessary to consider the way that these different aspects interact with each other. A key enabler in this technology is the intimate integration of optoelectronic devices with an ASIC substrate. In the following section, methods used for realizing OE-VLSI chips and associated packaging used in system demonstrators are described. For example, the appropriate design guidelines employed in the construction of some specific devices are articulated.

4.3 Heterogeneous Integration

Heterogeneous integration is used in order to achieve high density optical I/O. When doing so, it is important to assure compatibility of ASIC metals with the metals used in the flip-chip bump bonding. A related issue is the need to insure alignment of ASIC connection points with III-V device contact points. In the following sections, a heterogeneous integration strategy based on flip-chip bump bonding is summarized. To achieve the OE-VLSI ASICs described, 2D arrays of VCSELs and PDs were fabricated on separate substrates and subsequently integrated onto the ASIC die. In order to support the compact high-density micro-optical interconnects described in Figure 4.3, the VCSELs and PDs were interleaved. The next section describes: the design and target operating properties of the VCSELs and PDs, the OE device layout geometries, and heterogeneous integration techniques including flip-chip bonding and substrate removal of the interdigitated OE devices.

4.3.1 VCSEL and PD Design and Specifications

Table 4.1 summarizes the suitable optical and electrical properties of VCSELs and PDs. All property values are measured quantities except those stated as exact numbers, which are quoted from device manufacturer data sheets. As indicated in Table 4.1, the VCSELs used were designed to operate at 850 nm with threshold currents of 1.0 to 4.5 mA and slope efficiencies of 0.25 to 0.35 mW/mA. The devices were also designed to be backside-emitting because of the desire to flip-chip bond them to ASIC driver circuits, as described below. This necessitated removal of the GaAs substrate to minimize absorption of light. To achieve these objectives, VCSELs were fabricated with both the n contact and the p contact located on the top surface of the wafer to facilitate electrical contact to the ASIC circuits. Figure 4.4 shows a schematic of the VCSEL geometry indicating emission direction after substrate removal and integration to the ASIC. The p contact was formed above the top distributed Bragg mirror (DBR) and the n contact was brought to the substrate surface through mesa isolation and ion implantation. Figure 4.5 is a photomicrograph of a 2×2 VCSEL subarray before flip-chip bump bonding to a CMOS ASIC. The p and n contacts are indicated. The p and n contacts and the VCSEL active area are on a 125-μm horizontal

TABLE 4.1. PD and VCSEL optical and electrical properities

PD parameter	Symbol	Value
Responsivity	R	0.317A/W ± 10.8 mA/W
Junction capacitance	C_J	500 fF
Dark current	I_{dark}	6 nA

VCSEL parameter	Symbol	Value
Threshold current (@25°C)	$I_{TH}\|_{25°C}$	1.400 mA ± 75 μA
Slope efficiency (@25°C)	$\eta\|_{25°C}$	0.340 W/A ± 10 mW/A
I_{TH} temperature dependence	$\partial I_{TH}/\partial T$	15.1 μA°C
η temperature dependence	$\partial \eta/\partial T$	−2.2 mW/A°C
Differential resistance (2-5 mA)	$\partial R\|_{2-5mA}$	71.7Ω ± 12.1Ω
Threshold voltage (@2mA)	$V_{TH}\|_{2mA}$	1.580V ± 17mV
Junction capacitance	C_J	500 fF

and vertical pitch. Once bonded to the ASIC as per the description given below, the n contact and the DBR becomes the top (emitting) surface of the VCSEL.

The PDs were fabricated with square active areas of 50 μm on a side and 15 × 15 μm pads for the p and n contacts. The PD active areas were placed on a 125-μm vertical pitch. Figure 4.6 is a photograph of a 2 × 2 PD subarray before flip-chip bump bonding to a CMOS ASIC. The p and n contacts and a unit cell are indicated.

4.3.2 Heterogeneous Integration and Substrate Removal

The VCSELs and PDs were integrated onto the CMOS drivers using flip-chip bonding and substrate removal techniques. The contact areas on the

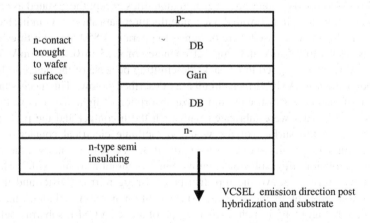

FIGURE 4.4. Schematic of the VCSEL geometry indicating the emission direction post substrate removal.

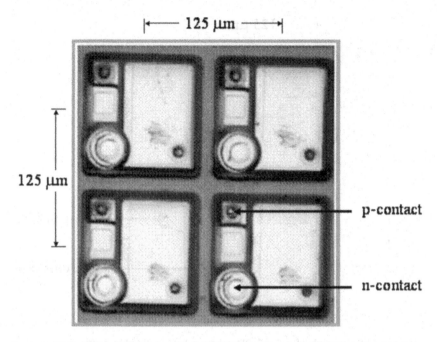

FIGURE 4.5. Photomicrograph of a 2x2 VCSEL subarray before after flip-chip bump bonding to a CMOS ASIC. The p and n contacts are indicated. The p and n contacts and the VCSEL active area are on a 125-μm horizontal and vertical pitch.

CMOS die for the VCSEL driver and PD receiver were identical to those on the optoelectronic devices. Heterogeneous integration was accomplished by employing relatively conventional photolithographic processes to deposit and lift off contact metals on the wafers followed by a precision assembly process using a flip-chip bonding tool. In the photolithographic step, a photoresist polymer was first spun out on the wafer and printed with the contact metal pattern and then developed. Indium was then evaporated onto the wafer and the photoresist was lifted off, leaving metal on the contact pads. This process was used for the VCSEL and PD wafers and the CMOS dies. The individual OE device dies were separated by mechanical dicing and then integrated onto a CMOS die using the precision alignment hybridization tool. The VCSEL die was first attached to the CMOS chip followed by dry etching to remove the substrate; integration of the PD die was accomplished next followed by substrate removal. The bonding of the indium metal contacts on the CMOS chip and on the OE devices was accomplished through a combination of force and controlled temperature.

In summary, the process resulted in electrical isolation of individual OE devices and allowed the interleaving of the VCSEL and PD devices onto a single CMOS die. Although individual dies were used to assemble this generation of OE-VLSI ASICs, migration of the process to the wafer level is relatively straightforward. The above process was found to be very effective in realizing the OE-VLSI ASICS used in the systems to date. As suggested, the process

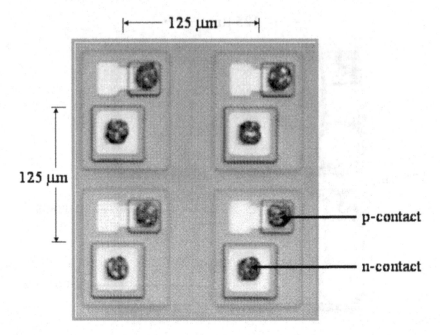

FIGURE 4.6. Photograph of a 2×2 PD subarray before flip-chip bump bonding to a CMOS ASIC. The p and n contacts and a unit cell are indicated.

is relatively straightforward and can be extended to wafer-level integration of devices with wafers of OE devices. The process lends itself to high levels of integration, thus permitting large optical I/O counts.

4.4 OE-VLSI ASIC Architectures

The architecture of an OE-VLSI ASIC largely determines the efficiency with which a specific application can be implemented. Determining the appropriate architecture for an application depends on requirements such as the complexity of the processing that must be performed on data or the size of the data set to be operated on. Each architecture type has strengths and weaknesses in terms of compatibility with automated place and route software tools used to realize the ASIC.

The three major architectures that have been employed in the design of OE-VLSI ASICs are the pixelized architecture, the modular architecture, and the area-distributed optical padframe architecture [11]. The first two are discussed in the next two sections, through examples of specific OE-VLSI ASICs.

4.4.1 Pixelized/Smart Pixel Architecture

The pixelized or smart pixel architecture is based on the development of a self contained unit cell comprised of an optical transmitter, optical receiver, and processing electronics that operate on the locally available data. Each unit cell is self-contained and does not necessarily communicate with other unit cells. The unit cell is tiled to form a 2D array of processing elements as illustrated in Figure 4.7a. The processing complexity that can be implemented in the unit cell is determined by the pitch of the OE Device (OED) arrays used in their construction, and the amount of space occupied by the transmitter and receiver circuits.

The pixelized architecture is compatible with applications that involve minimal data processing, such as control and routing functions [23, 24], optical backplanes [19], or multichip interconnection fabrics [7]. It is also compatible with applications that involve large data sets and simple instruction multiple data (SIMD) processing, which is commonly found in image processing applications [17, 21], or in binary neural associative memory applications [5]. The pixelized architecture is not suitable for applications that involve bit-parallel datapath architectures, such as microprocessor and digital signal processing (DSP) cores or random-access memory (RAM), in which highly regular row and column structures are used to achieve timing uniformity and to minimize routing congestion. Adapting these applications to a pixelized architecture would be highly inefficient from a design and layout perspective [11].

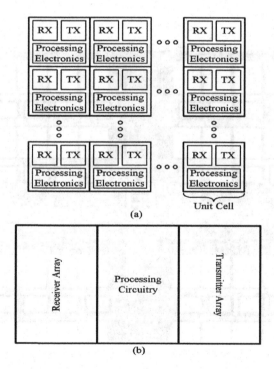

FIGURE 4.7. (a) The unit cell is tiled to form a 2D array of processing elements. (b) A modular layout methodology is shown in which the core processing circuitry and the optical transmitter and receiver arrays are physically separate modules.

The OE-VLSI ASIC required to support the compact point-to-point optical interconnect system described in Figure 4.3 was one employing the pixelized architecture. The interconnect requirements necessitated interleaving the VCSELs and PDs in a clustered geometry as shown in Figure 4.8. VCSELs and PDs were grouped together in clusters such that within each cluster, rows of VCSELs and PDs were interleaved. Resulting clusters consisted of eight VCSELs and eight PDs arranged in four rows. The pitch of the optoelectronic devices was 125 μm in both the horizontal and vertical directions; therefore, the VCSELs and PDs were on 125 μm horizontal by 250 μm vertical pitch. The complete 256-VCSEL and 256-PD array consisted of 32 clusters arranged in eight rows and four columns. The center-to-center spacing of clusters was 750 μm horizontally and 750 μm vertically. Figure 4.9 is a photograph of the complete OE-VLSI ASIC after VCSEL and PD integration. The CMOS device was a 10-mm × 10-mm chip fabricated in 0.35-μm technology with three metal layers. Figure 4.10(a) shows four clusters with 32 VCSELs forward biased (below threshold), and Figure 4.10(b) shows the entire VCSEL array biased above threshold. The distortion seen is caused by the optical system used to image the 3-mm × 6-mm VCSEL array.

FIGURE 4.8. Four clusters after flip-chip bump bonding and substrate removal. The VCSEL device was 110 μm × 112.5 μm and had a 20-μm-diameter active region. The PIN was 125 μm by 90 μm and had a 50- × 50-μm active region.

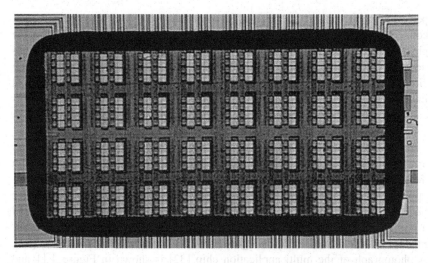

FIGURE 4.9. Photograph of the complete 256-element VCSEL and PD arrays integrated onto a CMOS ASIC.

4.4.2 Modular Architecture

In the design of a large-scale OE-VLSI ASICs, a modular layout methodology as shown in Figure 4.7b has been used, in which the core processing circuitry and the optical transmitter and receiver arrays are physically separate modules that can be separately placed and optimized [11, 32]. Wide and parallel on-chip electrical interconnects are then used to connect the different modules together. The OEDs reside with their respective transmitter and receiver circuits, and the pitch of the OED arrays determines the pitch of the circuits within the transmitter and receiver arrays, and correspondingly the allowed circuit complexity. A limiting

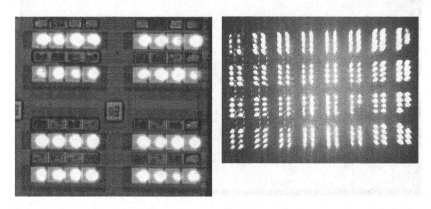

FIGURE 4.10. (a) Four clusters with 32 VCSELs forward biased (below threshold). (b) The entire VCSEL array biased above threshold. The distortion seen is caused by the optical system used to image the 3-mm × 6-mm VCSEL array.

performance factor for the modular layout methodology is the electrical interconnect used to connect the different modules of the chip. Depending on the physical size of the transmitter and receiver array, these electrical interconnections may be long, and this is of particular concern for the receivers. Buffering circuitry can be used to minimize the latency of driving long on-chip electrical interconnects at a target data rate. Additionally, introducing pipelining stages can be used to shorten the effective latency of long electrical interconnects in synchronous applications that employ the modular architecture [33].

By isolating the digital processing circuitry from the transmitter and receiver arrays, any desired level of processing complexity can be implemented with the modular architecture. The modular architecture has been used in the development of fully interconnected switches [18, 19], a crossbar switch [14], a page buffer chip [11]. an SRAM chip [25], a Fast Fourier Transformer [25], and a multiapplication chip that includes first in, first out (FIFO) buffers, pseudo-random bit sequence (PRBS) data generation, and forward error correction [32]. A photograph of the multi application chip [32] is shown in Figure 4.11 and highlights the principal sections: receiver array, digital section, and transmitter array. The 14.6-mm × 7.5-mm chip was built in a 0.25-μm, five-metal, single-poly, n-well foundry CMOS process. The VCSELs and PD were arranged in a 250-μm × 250-μm grid having 34 rows and 35 columns each. Of the 1190 devices in each OED array, 1080 were used with transmitter or receiver circuits. The remaining 110 OEDs in each array were physically present but not electrically connected to transceiver circuits.

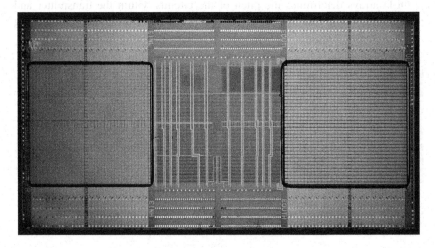

FIGURE 4.11. Photograph of the chip after flip-chip bump bonding of optoelectronic devices to an underlying CMOS device. The principal sections are shown: receiver array, digital section, and transmitter array.

4.5 Enabling Analog Circuit Designs and Performance

There are numerous possibilities with respect to transceiver design for OE-VLSI technology. Both single ended and differential designs have been employed. In the following section, both topologies are described including the merits of each. The discussion is partitioned into laser drivers and receivers.

4.5.1 Laser Drivers

This subsection presents an optical transmitter based on a low-cost 180-nm CMOS driver and an 850-nm VCSEL operating at 5 Gbps with a total power consumption of 18.35 mW [29]. A detailed bias-dependent modeling approach is used for the VCSEL and an original dual-power rail layout for reduced power consumption is introduced. A description of the simulation bench and test setup is included. The lowest reported power consumption per bandwidth in 180-nm CMOS, of 3.67 mW/Gbps, is achieved with an optical modulation amplitude (OMA) of 3 dBm. This subsection is organized as follows. The design and implementation of the optical transmitter are first outlined. Architectural design decisions are summarized, followed by a unique VCSEL modeling approach based on experimental data. The optical transmitter design, simulation, and implementation are then described. Finally, the test setup and measured optical transmitter performance are presented.

4.5.1.1 Architecture

Since applications with potentially wide deployment are contemplated, low manufacturing and operating costs are key drivers. By combining a 180-nm CMOS transmitter (TX) with an 850-nm VCSEL die, minimizing manufacturing costs was sought. In addition, CMOS is ideal for integrating digital processing functions with the transmitter on the same die, further reducing costs. Chip area utilization was minimized for the same reason, but also to fit parallel applications where VCSEL arrays have pitches equaling 125–250 μm [34]. As a result, on-chip inductors and cib-ohm resistors were avoided. Minimizing costs also implies minimizing power consumption. This was accomplished by proposing a novel double power-rail ring layout, which will be outlined later in this section. A compact three-stage transmitter architecture is used to keep power requirements low, below the target of 20 mW. As will be shown below, this yields the highest data rate over power consumption ratio reported to date in a 180-nm CMOS. Finally, high data rate and noise robustness influenced most of the architectural design decisions. Maximum performance was achieved through the use of NMOS-based circuits wherever possible. Because PMOS devices are much slower than NMOS transistors, they were used exclusively as loads. In addition, a detailed transition frequency (f_T) analysis was conducted on the available technology to determine the optimum performance bias point. Although this analysis generally applies to small-signal circuits, operating near f_T at DC insured operation in a region of optimum performance. Electrically differential signaling was used to increase the noise robustness of the transmitter. In addition, a current-steering driver was used. This topology generates much less switching noise than a current-switching driver [34].

4.5.1.2 VCSEL and Other Models

Accurate electrical simulations required a good VCSEL model. Reflection parameter (S11) measurements were conducted on several Emcore Gigalase™ VCSEL dies using a radio-frequency (RF) probing station at four different bias currents. The measured S11 curves at the four bias currents for three different dies are given in Figure 4.12. In addition, the differential resistance around the four bias conditions was calculated from the $V-I$ curves. The 2-mA bias current model was used for simulations of the optical transmitter design.

An electrical model topology derived from common approaches was used and is shown in Figure 4.13 [3, 35]. The VCSEL model parameters embody pad (RP, CP), junction (RJ, CJ), and mirror (RN, RM) characteristics. Measurements were imported in an Agilent ADS program and model parameters were fitted to the measured S11 and differential resistance independently for the four bias conditions. Optimization of the model parameters was conducted using both random and quasi-Newton search methods with the least-squares error function. The differential resistance at the four bias conditions, compared to the measured values, is given in Table 4.2. Excellent agreement between the measured and

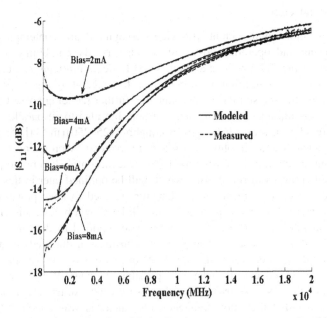

FIGURE 4.12. Experimental and modeled reflection parameter curves at four bias current levels.

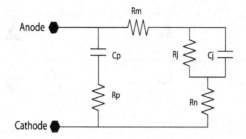

FIGURE 4.13. VCSEL electrical model topology.

TABLE 4.2. Experimental and modeled
VCSEL differential resistance at four
bias current levels

Bias current (mA)	δR_{DC}	
	Experimental (Ω)	Modeled (Ω)
2	103.4	104.0
4	83.3	83.0
6	73.1	73.0
8	67.0	67.0

modeled data is demonstrated. Additional models for the bond wires, test fixture, and package were included in a system simulation bench in Cadence Virtuoso™ with the extracted transmitter layout.

4.5.1.3 Three-Stage Transmitter Design, Implementation, and Simulation

A three-stage optical transmitter was used and it is illustrated in Figure 4.14 [27]. It comprises a differential preamplifier, a pulse-shaping stage, and a current-steering driver. The preamplifier was implemented as a common-gate differential stage, with a 100-Ω matching resistance between its inputs. A 0.4-V low-voltage differential signal (LVDS) is fed to the inputs and amplified near the VDD (1.8 V) and ground rails. The gain of the preamplifier stage is 11 dB and the 3-dB bandwidth for such an input signal is 6 GHz.

The pulse-shaping stage is designed as a variable-bias common-source differential pair. The bias of the pulse-shaping stage is varied to adjust the low voltage presented to the current driver. By presenting a low voltage at the current driver inputs that is slightly higher than V_X (see Figure 4.14), ringing at the output of the driver stage is significantly reduced [26]. This dynamic adjustment makes it possible to use a wide variety of modulation currents without degradation in performance.

The current driver steers a modulation current adjustable in the range 0–4 mA between its two differential pair branches. A bias current source is also included

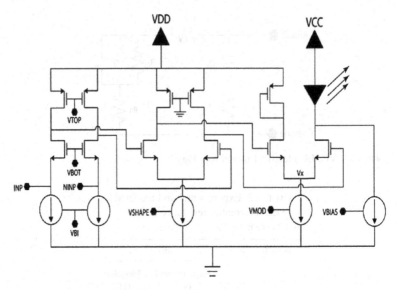

FIGURE 4.14. Three-stage optical transmitter schematic.

on-chip and is adjustable between 0 and 8 mA. As an optically single-ended approach was used, a dummy load mimicking the electrical behavior of a VCSEL was implemented in one branch as diode-connected PMOS transistors. Since the VCSEL forward voltage of 2 V is higher than VDD, the VCSEL was connected to a second power rail VCC of 3.3 V. Having the entire optical transmitter connected to VCC is the common approach [16]. On the other hand, this significantly increases power consumption and reduces performance since thick oxide transistors must be used.

A dual power rail layout was used where the two are implemented as rings around the die separated by a ground ring. The exterior ring is VCC and brings power to the VCSEL die exclusively. This original approach led to very low power consumption numbers. Having different power supplies for the two branches of the current-steering driver can induce additional switching noise on the power rails. This unbalance was mitigated by laying out a large coupling capacitor on-chip to increase the quality of the local power rails.

The optical transmitter extracted layout, combined with the VCSEL and test fixture models, was simulated with SpectreS at 55°C. As the VCSEL is to be uncooled during experiments, it is reasonable to assume the temperature of the CMOS and laser will rise significantly above room temperature. Simulations and experimental measurements showed that the test fixture used was too slow for operation above 2.5 Gbps. As a result, the LVDS inputs are applied directly to the TX inputs through an ideal 50-Ω transmission line. This represents an experimental setup where the input pads are directly probed with 50-Ω high-speed probes. A 2^{10}-1 pseudo-random bit sequence (PRBS) was used to obtain simulated eye diagrams at 2.5 and 5 Gbps shown in Figure 4.15. A modulation

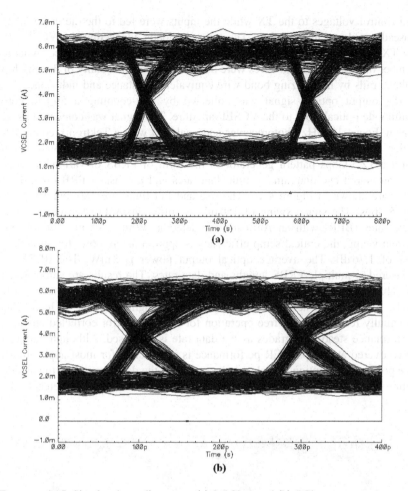

FIGURE 4.15. Simulated eye diagram at (a) 2.5 Gbps and (b) 5 Gbps.

current of 4 mA was applied to the VCSEL model, over a bias current of 2 mA. The threshold current and slope efficiency of the VCSEL are respectively 1 mA and 0.5 mW/mA. Current rise and fall times are 120 and 110 ps at 5 Gbps. The total simulated power consumption from the two voltage rails equals 21.75 mW, including 16.8 mW from the VCSEL modulation and bias currents. The power consumption due to the input ports is negligible.

4.5.1.4 Test Setup and Measured Performance

The optical TX die was fabricated by Taiwan Semiconductor Manufacturing Corporation (TSMC) through the Canadian Microelectronics Corporation. It was packaged with the Emcore VCSEL singlet in an 80-pin ceramic quad-flat package (CQFP) and mounted on a printed-circuit board (PCB). The PCB supplied power and control voltages to the TX while the inputs were fed to the die using 50-Ω Cascade coplanar microprobes. The two wirebonds connecting the VCSEL to the TX die were kept very short (200 μm) for maximum performance. A large number of power rail bond wires were used to increase the quality of the on-chip voltage rails by minimizing bond wire equivalent resistance and inductance.

The output optical signal was collected by butt-coupling a 50-μm core multimode optical fiber to the VCSEL aperture. The signal was converted using a New Focus 12-GHz photoreceiver and analyzed with a Tektronix communication signal analyzer. The LVDS 0.4-V input signals were generated by an Anritsu 12.5-Gbps pattern generator in the form of PRBS of various lengths. The measured eye diagrams at four data rates and a constant PRBS length of 2^{31}-1 are shown in Figure 4.16. The rise and fall times are 180 and 160 ps at 3.125 Gbps. The signal swing of 150 mV is sufficient to directly measure bit-error rates (BER) with an Anritsu error detector. With a 3.5-mA modulation current swing, the optical setup efficiency is approximately 66%, for an optical loss of 1.76 dB. The average optical output power is 2 mW. The BER was measured at multiple PRBS lengths and data rates. The results are summarized in Table 4.3. Errorless operation is achieved at a 3.125 Gbps data rate. A BER of 1×10^{-10} is still achieved with a maximum length PRBS at 4 Gbps. This essentially results in error-free operation for a forward-error corrected channel. Performance steadily degrades as the data rate is increased. Although data can be recovered at 5 Gbps, BER performance is insufficient for most applications. The power consumption, which is independent of the data rate due to the analog nature of the optical transmitter, was measured with precision multimeters as the

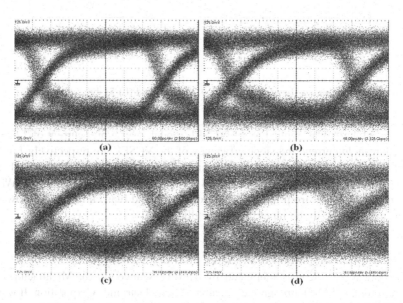

FIGURE 4.16. Measured eye diagram after transmission and conversion at (a) 2.5 Gbps, (b) 3.125 Gbps, (c) 4 Gbps, and (d) 5 Gbps.

TABLE 4.3. Measured BER at various data rates and
PRBS lengths

Data rate (Gbps)	PRBS length	Bit-error rate
1.250	2^{31}-1	Error free
2.500	2^{31}-1	Error free
3.125	2^{31}-1	Error free
4.000	2^{31}-1	$<1 \times 10^{-10}$
4.000	2^{15}-1	$<1 \times 10^{-11}$
4.000	2^{7}-1	$<1 \times 10^{-12}$
5.000	2^{15}-1	$<1 \times 10^{-7}$
5.000	2^{7}-1	$<1 \times 10^{-10}$

sum of the two power rail average currents. The total power consumption was
measured at 18.35 mW, composed of 13.35 mW of VCSEL power and 5 mW
of optical TX power. The power consumption with respect to simulations was
reduced by lowering the bias current to a minimum.

4.5.1.5 Discussion

The data rate and power consumption figures reported here compare advanta-
geously with previously published work on VCSEL-based optical transmitters.
Table 4.4 gives an overview of previous work encountered in the literature
[1, 15, 16]. The versatile optical transmitter design presented in this section
consumes less power than all other designs by a significant margin. While some
transmitters achieve higher data rates, the design presented here scales easily to
other CMOS generations and the author is confident that 10-Gbps operation and
low power consumption can be achieved in CMOS 130 nm or CMOS 80 nm,
for example. By normalizing the power consumption by the bandwidth (mW
per Gbps), an accurate portrait of the performance of optical transmitter designs
can be obtained. Only the work described by Kromer [15] has lower power
consumption per bandwidth. On the other hand, that design uses a state-of-
the-art 80-nm CMOS process with a 1.1-V power supply and a low-threshold

TABLE 4.4. Comparison of optical transmitter characteristics with previously published
designs

	Kucharski et al. [16]	Kromer et al. [15]	Annen and Eitel [1]	This design
Technology	CMOS 130 nm	CMOS 80 mn	SiGe 350 nm	CMOS 180 nm
Power supply voltage (V)	2.5	1.1	3.3	1.8
Max. data rate (Gbps)	20	10	10	5
Power consumption (mW)	120	25	132	18.35
Power consumption per bandwidth (mW/Gbps)	5.99	2.5	13.16	3.67

VCSEL. The proposed design would yield similar figures when transposed to that technology. Additionally, using a mature technology like 180-nm CMOS results in low cost, a key driver for the potential applications.

The partially unsettled challenge resides in noise minimization. While CMOS technologies present numerous advantages in terms of cost and integration, their noise characteristics are worse than other technologies such as SiGe or silicon-on-insulator (SOI) CMOS [8]. In addition, chip-on-board packaging and a high-quality 10-Gbps PCB should have a significant positive effect on the noise and data rate performance of the optical transmitter, compared to the CQFP/FR-4 PCB combination available in the work presented here. Reaching errorless 5 Gbps would then be feasible. This design has multiple input ports, giving access to bias and modulation current levels, and to pulse-shaping adjustments. This could be an asset for implementing dynamic adjustment based on low frequency feedback from the receiver. Such an approach would account for aging mechanisms and optical path length differences, and would eliminate the need for complex automatic power control (APC) circuitry in the transmitter. Alternatively, integrating the proposed design in a parallel optical interconnect is easily achievable. The transmitter presented here is compact and has low susceptibility to power supply switching noise. Finally, the low power consumption mitigates power dissipation issues present in parallel optical interconnect applications.

4.5.2 Receivers

After having addressed laser drivers in the previous section, two receiver designs employing differential inputs are described below. One of the principal reasons that differential optical signaling was used is to accommodate the DC-coupled nature of the input photocurrent. Optically single-ended receivers have a fixed decision threshold and variations in the average input photocurrent across an array of receivers can cause severe operational problems in receiver groups that are commonly biased and/or controlled. In optically differential receivers, a fully differential preamplifier architecture with common-mode feedback (CMFB) circuitry [31] stabilizes the operating point and common-mode output voltage of the preamplifier in the face of variations in common-mode input photocurrent.

Two preamplifier designs were implemented. One is based on a feedback-free common-gate amplifier (CGA) with a diode-connected load, and the other is based on a conventional differential transimpedance amplifier (TIA) with resistive feedback. Transistor-level schematics of the CGA and TIA are shown in Figure 4.17a and b, respectively. A common feature of the two preamplifier designs was the inclusion of circuitry (transistors MP in Figure 4.17) to allow for functional circuit testing prior to heterogeneous OED integration. After heterogeneous integration, the PDs would appear in parallel with transistors MP, as indicated by the dotted lines in Figure 4.17. The corresponding active-low digital control inputs nV_{tl} and nV_{tr} for the receivers in all common-control groups that form a digital-functional channel are connected together, and can be used to

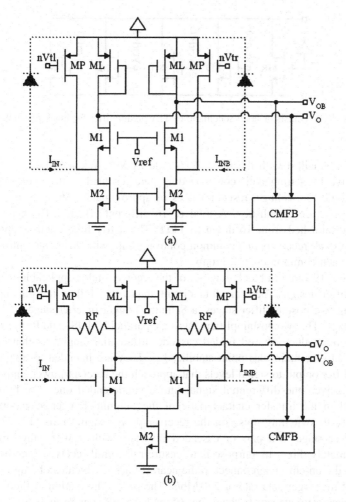

FIGURE 4.17. Transistor-level schematics for the (a) CGA and (b) TIA preamplifier circuits illustrating CMFB, test mode circuitry, and control inputs. The locations of the PDs after heterogeneous OED integration are indicated by dotted lines.

inject small amounts of current (approximately $60\,\mu A$) into either input of the preamplifier circuit, mimicking a differential input photocurrent.

The feedback resistances used in the TIA preamplifier configuration (resistors RF in Figure 4.17b) are implemented using active devices and can be tuned using digital control inputs as shown in Figure 4.18. One reference NMOS transistor (MR) and four other NMOS transistors (M0–M3) are all connected in parallel. MR was kept permanently conducting by having its gate terminal connected to the receiver supply voltage. This establishes a nominal resistance equal to approximately $16\,k\Omega$ for small current magnitudes. Transistors M0 through M3 have width-to-length (W/L) ratios progressively increasing by a factor of two.

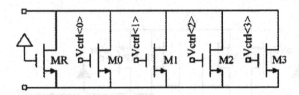

FIGURE 4.18. Transistor-level schematic of TIA preamplifier feedback resistance RF.

M0 is the smallest with the same W/L ratio as MR. When M0–M3 are made conductive by setting their corresponding gate terminal control voltages to a digital logic "1" voltage, resistor values of approximately 16, 8, 4, and $2\,k\Omega$, respectively, are established. A total of 16 different effective RF magnitudes can be established from 16 down to $1\,k\Omega$. The active-high control inputs are common to all receivers in a common-control group, with different control input sets for each common-control group.

Figure 4.19 shows a block diagram of the receiver and is applicable for either preamplifier design. There are four circuit stages that follow the preamplifier, including two postamplifier stages, a Schmitt–Trigger inverter stage, and a line driver stage. The two postamplifier stages are parallel NMOS– and P-type metal–oxide–semiconductor-based folded cascode differential amplifiers that employ feedback to maintain bias point stability [2]. They are intended to amplify the preamplifier output to signal levels that approach the receiver power supply rails and to convert the differential signal into a single-ended one. The hysteresis provided in the transfer characteristic of the Schmitt–Trigger stage provides immunity to switching noise on the receiver power supply rails [12, 23]. The line driver stage is a pair of cascaded inverters with a W/L stage ratio of approximately three. Its purpose is to ensure rail-to-rail receiver operation and to drive the on-chip interconnect to the nearest buffer, which can be up to $2\,mm$ away, at the target data rate of $250\,Mb/s$. The power dissipation of the receiver circuit was estimated to be between 8.5 and $9.5\,mW$ per receiver under most operating conditions at a data rate of $250\,Mb/s$.

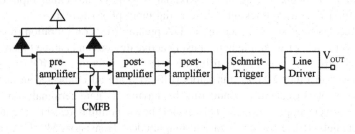

FIGURE 4.19. Receiver block diagram, applicable to either preamplifier design.

4.5.3 Discussion

Through the design, fabrication, and testing of these two large-scale OE-VLSI ASICs and numerous test chips in which different receiver and transmitter topologies were employed, a number of core conclusions are drawn on analytical, experimental, and simulation-based work. First, there needs to be some method of controlling the key set point parameters of the laser drivers and receivers across a subset of the entire array. This leads to higher operational yield. Second, the use of an optically and electrically differential architecture for the receiver and transmitter designs is optimal for OE-VLSI applications. Through his work, the author has found that optically and electrically differential architectures facilitate or optimize the implementation of several critical aspects of OE-VLSI ASIC design, including: (1) design for testability concepts and implementation for receiver and transmitter circuits; (2) the receiver and transmitter circuit generation of, and immunity to, switching noise on the voltage supply and ground rails and through the substrate; (3) the improvement of operational yield (percentage of functional circuits in a group of circuits that meet bit-error and data rate targets) in common bias and control receiver groups; and (4) the reduction of intrachannel receiver skew in parallel digital synchronous OE-VLSI applications and the reduction of individual receiver latency.

4.6 Conclusion

This chapter has described design rules, fabrication steps, and the performance of a number of successful optoelectronic-VLSI devices that have recently been demonstrated. Optically interconnected chips, boards, and backplanes offer a potential solution to this anticipated electrical interconnect bottleneck. A common design assumption is that optics provides increased connectivity through parallelism. Specifically, parallel optical interconnects are capable of providing high-bandwidth communication links both within and between high-performance electronic systems. The concept of the direct sourcing and termination of optical signals on silicon has been proposed as a method to relieve the off-chip communication bottleneck. The objective of the approach is to create optical inputs and outputs (I/O) that operate in conjunction with traditional electrical I/O. We have shown examples where this partitioning is effective and have proven that significant progress has been made in the enabling technologies required to achieve this objective. Much of the reported progress is as a result of a series of successful OE-VLSI ASIC demonstrations. Through the realization of these devices, research in a wide range of areas has been accomplished, including optoelectronic device design, electronic circuit design, and optical design and packaging.

Acknowledgments. This work was partially supported by the Natural Sciences and Engineering Council (NSERC) and industrial partners, through the Agile All-Photonic Networks (AAPN) research network, the Canadian Microelectronics

Corporation (CMC), and NSERC and FQRNT. The author gratefully acknowledges contributions from current and former members of the Photonic Systems Group at McGill University.

References

1. R. Annen and S. Eitel, Low power and low noise VCSEL driver chip for 10Gbps applications, in *Proc. 17th Annual Meeting of the IEEE LEOS,* pp. 312–313 (2004).
2. M. Bazes, Two novel fully complementary self-biased CMOS differential amplifiers, *IEEE J. Solid-State Circuits* **26**, 165–168 (1991).
3. P.R. Brusenback, S. Swirhun, T.K. Uchida, M. Kim, and C. Parsons, Equivalent circuit for vertical cavity top surface emitting lasers, *Electron. Lett.* **29**, 2037–2038 (1993).
4. M. Châteauneuf, A.G. Kirk, D.V. Plant, T. Yamamoto, J.D. Ahearn, and W. Luo, 512-channel vertical-cavity surface-emitting laser based free-space optical link, *Appl. Opt.* **41**, 5552-5561 (2002).
5. D. Fey, W. Erhard, M. Gruber, J. Jahns, H. Bartlet, G. Grimm, L. Hoppe, and S. Sinzinger, Optical interconnects for neural and reconfigurable VLSI architectures, *Proc. IEEE* **88**, 838–848 (2000).
6. K.M. Geib, K.D. Choquette, D.K. Serkland, A.A. Allerman, and T.W. Hargett, Fabrication and performance of two-dimensional matrix addressable arrays of integrated vertical-cavity lasers and resonant cavity photodetectors, *IEEE J. Sel. Top. Quantum Electron.*, **8**, 943–947 (2002).
7. M.W. Haney, M.P. Christensen, P. Milojkovic, G.J. Fokken, M. Vickberg, B.K. Gilbert, J. Rieve, J. Ekman, P. Chandramani, and F.E. Kiamilev, Description and evaluation of the *FAST-Net* smart pixel-based optical interconnection prototype, *Proc. IEEE* **88**, 819–828 (2000).
8. B. Jalali, S. Yegnanarayanan, T. Yoon, T. Yoshimoto, I. Rendina, and F. Coppinger, Advances in silicon-on-insulator optoelectronics, *IEEE J. Sel. Top. Quantum Electron* **4**, 938–947 (1998).
9. N.M. Jokerst, R. Beats, and K. Kobayashi (Eds.), *IEEE J. Sel. Top. Quantum Electron.* **5**, 143–398, (1999) and references therein.
10. N.M. Jokerst, and A.L. Lentine (Eds.), *IEEE J. Sel. Top. Quantum Electron.* **2**, 3–148, (1996) and references therein.
11. F.E. Kiamilev, and R.G. Rozier, Smart pixel IC layout methodology and its application to photonic page buffers, *Int. J. Optoelectron.* **11**, 199–215 (1997).
12. A.V. Krishnamoorthy et al., Triggered receivers for optoelectronic VLSI, *Electron. Lett.* **36**, 249–250 (2000).
13. A.V. Krishnamoorthy, L.M.F. Chirovsky, W.S. Hobson, R.E. Leibenguth, S. P. Hui, G.J. Zydzik, K.W. Goossen, J.D. Wynn, B.J. Tseng, J. Lopata, J.A. Walker, J.E. Cunningham, and L.A. D'Asaro, Vertical cavity surface emitting lasers flip-chip bonded to gigabit/s CMOS circuits, *IEEE Photon. Technol. Lett.* **11**, 128–130 (1999).
14. A.V. Krishnamoorthy, J.E. Ford, F.E. Kiamilev, R.G. Rozier, S. Hunsche, K.W. Goossen, B. Tseng, J.A. Walker, J.E. Cunningham, W.Y. Jan, and M.C. Nuss, The AMOEBA switch: An optoelectronic switch for multiprocessor networking using dense-WDM, *IEEE J. Sel. Top. Quantum Electron.* **5**, 261–275 (1999).

15. C. Kromer, G. Sialm, C. Berger, T. Morf, M. Schmatz, F. Ellinger, D. Erni, G.-L. Bona, and H. Jäckel, A 100mW 4x10Gb/s transceiver in 80nm CMOS for high-density optical interconnects, in *Proc. IEEE Int. Solid-State Circuits Conf.*, 18.4 (2005).

16. D. Kucharski, Y. Kwark, D. Kuchta, D. Guckenberger, K. Kornegay, M. Tan, C.-K. Lin, and A. Tandon, A 20Gb/s VCSEL driver with pre-emphasis and regulated output impedance in 0.13 μm CMOS, in *Proc. IEEE Int. Solid-State Circuits Conf.*, 12.2, pp. 5–7 (2005).

17. C.B. Kuznia, J.-M. Wu, C.-H. Chen, B. Hoanca, L. Cheng, A.G. Weber, and A.A. Sawchuk, Two-dimensional parallel pipeline smart pixel array cellular logic (SPARCL) processors-chip design and system implementation, *IEEE J. Sel. Top. Quantum Electron.* **5**, 376–386 (1999).

18. A.L. Lentine, K.W. Goossen, J.A. Walker, L.M.F. Chirovsky, L.A. D'Asaro, S.P. Hui, B.J. Tseng, R.E. Leibenguth, J.E. Cunningham, W.Y. Jan, J.-M. Kuo, D.W. Dahringer, D.P. Kossives, D.D. Bacon, G. Livescu, R.L. Morrison, R.A. Novotny, and D.B. Buchholz, High-speed optoelectronic VLSI switching chip with > 4000 optical I/O based on flip-chip bonding of MQW modulators and detectors to silicon CMOS, *IEEE J. Sel. Top. Quantum Electron.* **2**, 77–84 (1996).

19. A.L. Lentine, K.W. Goossen, J.A. Walker, J.E. Cunningham, W.Y. Jan, T.K. Woodward, A.V. Krishnamoorthy, B.J. Tseng, S.P. Hui, R.E. Leibenguth, L.M.F. Chirovsky, R.A. Novotny, D.B. Buchholz, and R. L. Morrison, Optoelectronic VLSI switching chip with greater than 1 Tbit/s potential optical I/O bandwidth, *Electron. Lett.* **33**, 894–895 (1997).

20. Y. Li, E. Towe, and M.W. Haney, (Eds.), *Proc. IEEE* **88**, 721–876 (2000) and references therein.

21. N. McArdle, M. Naruse, and M. Ishikawa, Optoelectronic parallel computing using optically interconnected pipelined processing arrays, *IEEE J. Sel. Top. Quantum Electron.* **5**, 250–260 (1999).

22. D.A.B. Miller (Ed.), *IEEE/OSA J. Lightwave Technol.* xx, xx–xx, (2003) and references therein. Vol 22, No.9, pp.2018–2222, 2004.

23. D.V. Plant, M.B. Venditti, E. Laprise, F. Faucher, K. Razavi, M. Chateauneuf, A.G. Kirk, and J.D. Ahearn, A 256 channel bi-directional optical interconnect using VCSELs and photodiodes on CMOS, *IEEE J. Lightwave Technol.* **19**, 1093–1103 (2001).

24. D.R. Rolston, D.V. Plant, T.H. Szymanski, H.S. Hinton, W.S. Hsiao, M.H. Ayliffe, D. Kabal, M.B. Venditti, A.V. Krishnamoorthy, K.W. Goossen, J.A. Walker, B. Tseng, S.P. Hui, J.C. Cunningham, and W.Y. Jan. A hybrid-SEED smart pixel array for a four-stage intelligent optical backplane demonstrator, *IEEE J. Sel. Top. Quantum Electron.* **2**, 97–105 (1996).

25. R.G. Rozier, F.E. Kiamilev, and A.V. Krishnamoorthy, Design and evaluation of a photonic FFT processor, *J. Parallel Distrib. Comput.* **41**, 131–136 (1997).

26. K.R. Shastri, K.N. Wong, and K.A. Yanushefski, 1.7 Gb/s NMOS laser driver, in *Proc. Custom Integrated Circuit Conf.*, pp. 5.1.1–5.1.4 (1988).

27. K.R. Shastri, K.A. Yanushefski, J.L. Hokanson, and M.J. Yanushefski, 4.0Gb/s NMOS laser driver, in *Proc. Custom Integrated Circuit Conf.*, pp.14.7/1–14.7/3 (1989).

28. SIA (Semiconductor Industry Association), *International Technology Roadmap for Semiconductors: 1999* Edition, International SEMATECH, Austin, TX (1999).

29. J.-P. Thibodeau, C. Murray, and D.V. Plant, An ultra low-power 5 Gbps VCSEL-based optical transmitter for optical interconnect applications and access networks, submitted.
30. H. Thienpont, (Ed.), *IEEE J. Sel. Top. Quantum Electron.* Vol.9, No.2, MARCH/APRIL, pp. 347–676, 2003.
31. P.M. Van Petegham, and J.F. Duque-Carillo, A general description of common-mode feedback in fully-differential amplifiers, *Proc. IEEE Int. Symp. Circuits Syst.* **4**, 3209–3212 (1990).
32. M.B. Venditti, E. Laprise, J. Faucher, P.-O. Laprise, J.E. Lugo, and D.V. Plant. Design and test of an Optoelectronic-VLSI (OE-VLSI) chip with 540 element receiver/transmitter arrays using differential optical signaling, *IEEE J. Sel. Top. Quantum Electron.*, **9**, 361–379 (2003).
33. M.B. Venditti, Receiver, transmitter, and ASIC design for Optoelectronic-VLSI applications, Ph.D. thesis, McGill University (2003).
34. M.B. Venditti, J.D. Schwartz, and D.V. Plant. Skew reduction for synchronous OE-VLSI receiver applications, *IEEE Photon. Technol. Lett.* **16** 1552–1554 (2003).
35. D. Wiedenmann, R. King, C. Jung, R. Jäger, R. Michalzik, P. Schnitzer, M. Kicherer, and K.J. Ebeling, Design and analysis of single-mode oxidized VCSEL's for high-speed optical interconnects, *IEEE J. Sel. Top. Quantum Electron.* **5**, 503–511 (1999).
36. J.W. Goodman, F.J. Leonberger, S.V. Kung, and R. Athale, Optical Interconnects for VLIS Systerms, Proceedings of the IEEE 72(**7**): 850–866, 1984.

5

Luminescence of Gold Nanoparticles

Luca Prodi, Gionata Battistini, Luisa Stella Dolci, Marco Montalti,
and Nelsi Zaccheroni

Department of Chemistry "G. Ciamician," University of Bologna,
via Selmi 2, 40126 Bologna Italy
luca.prodi@unibo.it
gionata.battistini@studio.unibo.it
luisastella.dolci2@unibo.it
marco.montalti2@unibo.it
nelsi.zaccheroni@unibo.it

5.1 Introduction

The field of nanotechnology deals with research and technology development in the length scale range of approximately 1 to 100 nm designing and using systems and devices characterized by novel properties rising from their size.

Research effort in nanotechnology has been enormous in the past few years, and it is continuously increasing. The U.S. National Science Foundation global estimates for 2004–2005 of the market of nanotechnology (or products containing nanotechnology) range from 8 to 68 billion dollars, and predict an increase for the same market up to 1 trillion dollars in the next 20 years [1].

More than 3 billion dollars was spent in 2003 on nanotechnology research, by hundreds of governments, universities, and private centers [2], but progresses in developing commercial nanotechnology applications have been and still are relatively slow. There are fears and uncertainties about the present and above all the future of nanotechnology that create difficulties for investors. Main concerns are about health and environmental aspects of nanoscale materials besides the sociological, ethical, and psychological impact that the introduction of this new technology will have. Michael Crichton's novel *Prey*, based on the pretended dangers of nanotechnology, was a best-seller in 2002 [3].

It is nevertheless clear that nanotechnology will have a great and fundamental impact on many sectors of human life in the near future and many publications testify to the importance of a rigorous and scientific approach and understanding of the situation to date [1, 2, 4].

One of the most valuable and unique characteristics of nanotechnology is its intrinsic interdisciplinarity: chemistry, physics, biology, materials science, medicine, and engineering are all involved and substantially contributing in the development of this field.

In this context, great and increasing interest is devoted to nanoparticles: particles with controlled dimensions on the order of nanometers. They can be metal or semiconductor aggregates, but also polymeric or mixed ones. They are usually described as small spheres with a radius in the range scale of nanometers but their shape can actually vary very much.

Applicative uses of nanotechnology are in general still very limited, but nanoparticles have already found many industrial applications in a very wide range of fields: electronics, optoelectronic, biomedical, pharmaceutical, cosmetic, catalytic, and materials areas, with products including chemical-mechanical polishing, magnetic recording tapes, sunscreens, automotive products, catalyst supports, biolabels, electroconductive coatings, and optical fibers. One of the cut-edge fields in nanoparticle research and new products is the enhancement of biological imaging for medical diagnostics, drug discovery and delivery.

Nanoparticles are extensively used also in catalysis being extremely reactive and ensuring a large surface area per volume unit, and research is very lively in this sector too, as testified by the large number of publications. This subject, even though very interesting, deals with an area of research beyond the subject of this chapter, and therefore it will not be considered here. Readers interested in this topic can examine interesting books and papers [5].

The impressive range of interests and applications of nanoparticles arises from the versatility and tunability of their properties. As predicted several years ago [6], properties of bulk materials change when small—down to a few atoms— aggregates are considered. The properties of little clusters sized in the range of 1–100 nm depend not only on the material but also on their size and shape, besides the environmental conditions that their surfaces experience. Moreover, many different capping agents can be used in general to derivatize their surface and the variety of chemical materials that can be used allows one to control and tailor their properties, such as solubility or affinity, and to introduce desirable functionalities, including receptors, reactive sites, electroactive or photoactive functionalities, DNA strains, and so on (Figure 5.1).

In this vast context, we focus our attention on gold nanoparticles showing luminescence properties, as we believe this area presents a particular appeal.

Among the different kinds of nanoparticles, gold ones are extensively studied for many reasons. First, their synthesis is straightforward, since many different reagents lead to reduction of salts of this noble metal; in addition, they are the most stable metal colloids and they present very peculiar electronic, magnetic, and optical properties related to their size due to the so-called "quantum size effect" (discussed later in this chapter).

Their colors have been known and used since antiquity (centuries before Christ). Colloidal gold was used to decorate polychromatic glasses and ceramics, assuring nice iridescent effect, and in medical treatments for a variety of pathologies (unstable mental and emotional states, arthritis, blood circulation, tuberculosis, and so on) [7] and in medical tests [8]. The first scientific attempt

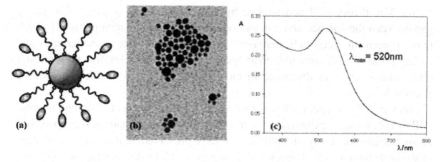

FIGURE 5.1. (a) Schematic representation of a capped nanoparticle. (b) TEM image of gold nanoparticles with an average diameter of 7 nm capped with alcanethiols. (c) Absorption spectra of the same nanoparticles of (b), circa 1×10^{-8} M in acetonitrile.

to describe and understand its chemical characteristics dates back to 1718, when Hans Heinrich Helcher published a complete treaty on colloidal gold [9]. More than one century later, in 1857, Faraday reported its synthesis via the reduction of a gold salt in aqueous solution and a study about its optical properties [10].

In recent decades, many chemical and physical methods for the preparation of gold and silver colloids have been proposed but the real breakthrough was introduced by Brust and Schiffrin in 1994 [11], which, for its ease, efficiency, and tenability, allowed significant advances in the field, as extensively reported in excellent books [12], reviews [13], and a large number of scientific papers (more than 10,000 in the last 12 months).

This research effort can be ascribed to the peculiar properties of these materials that are relevant for both fundamental and applied studies and that are induced by the "quantum size effect." Metal and semiconductor bulk properties can be explained referring to electrons that occupy energy bands, whereas in single atoms electrons are distributed in discrete energy levels. When only a few atoms are linked together forming very small clusters with dimensions in the range of a few nanometers, their electronic configuration lays somewhat in between; it dramatically depends on their size and can be fully described only by quantum mechanics. Therefore, this effect, induced by their small dimensions, is called the "quantum size effect". The electronic (optical, electrochemical, magnetic, and catalytic) properties of nanoparticles differ from those of bulk materials because their valence electrons experience different energies and distributions.

One of the most apparent and exploited characteristics is the peculiar color of metal (and semiconductor) nanoparticles, which is a consequence of the so-called "plasmon resonance band." This optical absorption band is typical of metal colloids, and shows a maximum in the spectral range between 500 and 600 nm, which confers to the particle a characteristic color that varies from red-ruby to blue-brownish, depending on the size, shape, and environmental polarity. The size-dependent plasmon resonance band has been theoretically described by the Mie theory [14] and studied by many authors. Its origin is the energy gap between the valence and conduction electrons in metal nanoparticles, which is not present in metal bulk materials. For particularly small nanoparticles and certain polarity environmental conditions, metal nanoparticles also present their own luminescence, as discussed in more depth at the end of this chapter (see Section 5.3).

From the electrochemical point of view, cyclic (CV) and differential pulse (DPV) voltammetry experiments with silver and gold nanoparticles, capped with alkanethiolate and fairly monodispersed, have presented an interesting redoxlike charging behavior [15]. Particles showing up to 15 peaks, corresponding to 15 different redox states, have been characterized (Figure 5.2) [16].

This means that single electron transitions are possible allowing the observation of the so-called "Coulomb blockade," if the stepwise charging energy ($E_C = e^2/2C$) is larger than the thermal one ($E_T = k_B T$, where k_B is the Boltzmann constant) [6, 17]. This has been observed for many samples using STM instrumentations and observing the electron tunneling between the tip and the nanoparticle [18].

All the characteristic properties described above have been observed and described by Schmid and co-workers, in particular for Au_{55} nanoparticles stabilized with phosphine ligands, with an exhaustive and extremely wide work that was started a few years ago [19].

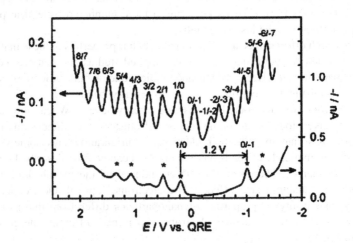

FIGURE 5.2. DPV responses for AuNP solutions measured at a Pt microelectrode; as-prepared 177 μM C6S-Au_{147} (upper) showing 15 high-resolution peaks relative to single electron - transfer events termed quantized double layer charging (QDL), and 170 μM C6S-Au_{38} (lower) showing a HOMO–LUMO gap. It can be seen that the as-prepared solution contains a residual fraction of Au_{38} that smears out the charging response in E regions where QDL peaks overlap. The electrode potential scanned negative to positive. (Reproduced from Ref. 16. Copyright 2003 American Chemical Society.)

Despite the research in this field being so dynamic and wide, a topic that has not been very deeply explored is the investigation of photophysical and photochemical properties of materials obtained by organizing specifically functionalized chomophores on the surface of metal colloids. This process can yield heteronanomaterials with a metal core and an organic shell that could be interesting for the design of new photoactive devices with a variety of possible applications (e.g., sensing, drug delivery, catalysis). The reason for the apparently limited interest in this area is the observation that metal nanoparticles generally quench the luminescence of many chromophores, via energy and/or electron transfer mechanisms. This is, however, not the end of the story, as is shown by the very interesting works described in this review. Before this discussion, however, we explain why luminescence is a valuable property from the point of view of both basic research and applications.

Luminescence spectroscopy is a powerful technique: fluorescence measurements are usually very sensitive (even single molecule detection is possible, although only under special conditions), low cost, easily performed, and versatile, offering subnanometer spatial resolution with submicrometer visualization and submillisecond temporal resolution. A wide number of parameters can be measured to get information on the system under study. Moreover, the variables of the system can be tuned to optimize the measurement in specific environments, such as the excitation and emission wavelengths, the time window for signal collection, the polarization of the excitation or the emission beams. Other properties, such as excited state lifetime and fluorescence anisotropy, which are less affected by the environmental conditions, also give valuable information.

Luminescence is the most straightforward property for molecular recognition and labeling, for medical, biological, environmental, and general analytical purposes. To achieve the best sensitivity in these applications, one should optimize some photophysical parameters: the system should present a high molar absorption coefficient (ε) and a high luminescence quantum yield (Φ). The luminescence intensity, in very diluted solutions, is in fact directly proportional to these quantities. It is evident that if one can organize a large number of chromophores in a single system, the value of at least ε will be enhanced. Moreover, in general, the organization of photophysically active units in structures such as derivative surfaces and conjugated polymers, typically gives rise to collective effects that can be exploited for the design of new functional materials. Among the different possibilities, the modification of nanoparticles as a suitable path to constrain a set of fluorescent units into an organized network is an intriguing example. Nanoparticles derivatized with fluorophores are, in fact, very promising for the design of labels, sensors, and other photoactive materials, due to the relative ease of their synthesis and mentioned properties [20].

We address our attention to nanoparticles that have gold cores and present luminescence properties. We consider both stabilized particles by species linked to their surface, and particles without capping material.

5.2 Chromophores Interacting with Gold Nanoparticles

5.2.1 How to Bind the Chromophores and Evidence the Binding Event

As mentioned, a large number of physical and chemical methods to synthesize gold nanoparticles are known [13 and references herein]. We will not deal here with the physical roots since we are interested, in this section, in discussing heterosystems that present chromophores directly linked at the surface of the gold nanoparticles. It is well established that many functionalities can bind to metallic gold, including sulfur-containing moieties such as thiols, thioethers, thioesters, disulfides, isothiocyanates, and then phosphines, amines, citrate, carboxylates, yielding, in all cases, a passivation of the surface that stabilizes nanoparticles preventing their coalescence. It has to be stressed that all these capping agents allow different degrees of stability. It is well known, in fact, that thiols form a covalent S-Au bond presenting the highest affinity for this metal; however, also among alkylated thiols there are differences, and hydrocarbon chains in the length range between C6 and C18 form the most stable structures.

Faraday was the first to report a scientific approach for gold colloid synthesis [10]. Chemical methods are generally based on the reduction of gold salts. The first quantitative, very easy, and still widely used method was published by Turkevitch in 1951 [21]. It yields gold nanoparticles stabilized with citrate and with an average diameter of 20 nm, presenting negative charges at the surface that make them soluble in water. Nevertheless, citrate moieties do not form very strong bonds with gold atoms and the material is not very stable. A great improvement was the introduction of a new method that allows the use of stronger ligands such as thiol-derivatized ones. This method was published by Brust and colleagues in 1994 [11], and it is probably still the most popular synthetic procedure. An AuIII complex is reduced after its phase transfer from water to organic media, in the presence of the thiol-ended stabilizing agent. This is an easy and versatile procedure that yields fairly monodispersed gold nanoparticles with average diameter that depends on the experimental conditions [22] and so stable that they can be handled as a normal chemical (dried, stored, and redispersed). The same procedure can be used to directly introduce other functionalities at the surface such as thioethers, thioesters, isothiocyanates, and amines provided that they are soluble in the organic solvent used for the synthesis. A further modification of the procedure included the preparation of clusters stabilized with weak ligands and then their exchange with stronger ones [23].

The first concern has always been the demonstration that the organic molecules used as capping materials are actually bound to the nanocluster surface. This can be verified in several ways. The passivating process, in fact, is based on the inter-action of the approaching molecules with the electrons of the external atoms of the nanoparticles, inducing rearrangements of their electron density. On successful binding, the NMR spectra of the molecules adsorbed on the surface are altered, revealing broadened picks, sometimes different chemical shifts [24], and, in the case of SH ligands, the total disappearance of the R-CH$_2$-S-Au proton signal [25].

Depending on the linking group, other effects have been revealed and studied, such as the interaction of the amine lone pair electrons with the metal or thioester and thiocarbonate hydrolysis, as discussed in the next section (Section 5.2.2).

Noncovalent bondings, such as electrostatic interactions or hydrogen bonds, also allow the efficient absorption of organic molecules on gold nanoparticles. Taking advantage also of the unique ability of nanoclusters to spontaneously form regular arrays when reasonably monodispersed, a large amount of recent research has been dedicated to surface-enhanced Raman scattering (SERS), another powerful spectroscopic technique, which allows the detection of the organic absorbed molecules in very low concentration, down to a single molecule.

The easiest ways to prove molecular binding to the surface remain the photophysical techniques, starting from UV-visible absorption measurements. In fact, the ligands on the surface usually perturb the position and shape of the surface plasmon band (SPB), which is determined by free electrons occupying the conduction band. Moreover, the SPB depends not only on material, dimension, and shape of the core of the nanoparticles, but also on the medium dielectric constant, refractive index, and charge [26]. The capping moieties shell around the nanoparticles is expected to modify all these quantities, causing a shift of the SPB with respect to uncapped clusters, and the changes are expected to be larger the stronger is the binding on the surface. Largest effects are observed with thiolated ligands or ligands that exhibit significant charge exchange with the nanoparticles.

When fluorophores are used as capping agents, spectrofluorimetry can be additionally exploited to gain useful information on the system or for applications. Luminescent bound molecules can experience new excited state deactivation possibilities such as energy and/or electron transfer with the metal and changes in both radiative and nonradiative decay rates. Up to quite recently, only quenching effects of the metal nanocore on the surface-bound chromophores had been reported causing a general impression that these heteromaterials were not suitable for applications in photodevices or biophotonics. However, recently, a few papers have been published, which describe luminescent chromophore bound to nanoparticles. In the next sections we provide an overview of the state of the art regarding this topic.

5.2.2 Quenching of Fluorophores Directly Linked to Nanoparticles

To our knowledge, the first paper reporting covalent fuctionalization of gold nanoparticles with a chromophore was published in 1997 [27], but no spectroscopic study was presented until Murray and co-workers [28] reacted 5-(aminoacetamido)-fluorescein with tiopronin monolayer-protected gold nanoparticles to yield an organic–inorganic system with an average number of fluorophores per nanoparticle equal to 3.7 units.

Specttrofluorimetric measurements revealed a strongly quenched emission from the fluorescein moiety but the luminescence band maintained the same shape and position as for the unbound chromophore. Because a reference experiment ruled out the possibility of quenching or self-absorption at those concentrations, this was the first experimental observation of the quenching effect due to the interaction with a gold nanoparticle of a chromophore covalently linked to the surface through its thiol derivative. After that pioneering work, many studies have been dedicated to the clarification of the reasons and mechanisms of this quenching effect and to this aim many different chromophores have been linked to the surface of gold nanoclusters such as dansyl, stilbene, porphyrins, fullerene, pyrene, and cyanine as discussed below.

Shortly after, the same authors prepared gold nanoparticles with an average of 140 Au atoms per particle, stabilized with a mixed monolayer of alkanethiolate and alkanethiolate derivatized with dansyl groups [29]. Varying (1) the number of chromophores per nanoparticle, (2) the length of the alkyl linker, and (3) the ratio length alkanethiolate/derivatized alkanethiolate on the same nanoparticle, their experimental results revealed in all cases a quenching of the dansyl emission. As expected, the energy transfer was less efficient when the distance from the luminescent unit and the gold core was longer but, since the linker was flexible, this depended on all three variables mentioned above. When the fluorophore is at the end of a long linker chain and the surrounding alkanethiolate monolayer is shorter, it experiences a large thermal mobility freedom that causes folding and levels the effect of the long spacer, suggesting a through-space quenching. The loading dependency of the emission and in particular lower quenching efficiency for higher loadings, was rationalized as reflecting an increased steric crowding and therefore a minor thermal fluctuation.

The same explanation was suggested for the reduced energy transfer efficiency for high percentage of ω-fluorenyl-alkane-1-thiolate on the surface of gold nanoparticles with an average diameter of 2 nm [30]. Fox and co-workers have also performed steady-state and time-resolved fluorescence studies that demonstrate that the quenching of the fluorescence of this system (to 5% of an equivalent one in which the fluorophores are free in solution) is due neither to a competitive absorption by the gold-monolayer-protected cluster nor to an enhanced intersystem crossing efficiency, but to a quick deactivation of the excitation energy through electronic coupling of the excited fluorenyl group with the gold nanoparticle. The possibility that the emission quenching was due to an electron transfer pathway was also ruled out, since no evidence of anion or cation radical transient absorption was found for the system at any of the tested wavelengths.

The same authors have also investigated an analogous system, obtained using 6-thiohexyl-3-nitro-4-(4'-stilbenoxymethyl)-benzoate to passivate the surface of Au_{500} clusters [31]. This moiety presents three very interesting and different features: it can undergo photoisomerization, photocleavage, and [2+2] photodimerization. The geometric isomerization of the stilbene and the photodeprotection of the phenolic group proved to be still active when the capping agent was linked on the metal core, even though the selective activation of them was possible only via triplet sensitization of the stilbene group. Photodimerization of the *trans*-stilbene, instead, was blocked suggesting that the packing of the units on the nanoparticle surface is not very dense or at least looser than in monolayer films on gold flat surfaces where this process is present. Even more interestingly, the luminescence of the system is strongly quenched as in the parent ones that present *trans*-4-(mercaptoheptoxy)stilbene at the surface and different core diameters ranging from 1.4 to 5.2 nm [32]. The fluorescence quenching depends on the particle size, being more efficient the smaller the metal core. The authors suggest that this depends on the higher conformational freedom of the stilbene moieties on small metal clusters in comparison with the densely packed ones on bigger clusters. Their flexibility allows folding that takes them closer to the Au surface yielding a more efficient through-space energy transfer.

All of these results agree on the importance that the distance between the chromophore and the metal plays in these systems. A recent paper proposes a very interesting study of the changes in radiative and nonradiative molecular decay rates as a function of this parameter [33]. The authors synthesized gold nanoparticles with an average diameter of 12 nm capped with different amounts of single-stranded oligonucleotides (ssDNA) modified with an SH group at one end and a Cy5 dye at the other end. ssDNA as a spacer allows one to obtain well-defined chromophore–metal surface distances, ranging between 2 and 16 nm, by changing the number of oligonucleotides and the surface loading (Figure 5.3). After neglecting electron transfer as a route of fluorescence quenching, they determined via time-resolved photoluminescence spectroscopy the values of radiative and nonradiative decay rates of the fluorescence that was suppressed by 95% in the most quenched system. The photoluminescence quantum efficiency is distance dependent, and the radiative rate decreases monotonically with decreasing distance while the nonradiative one increases although much weaker than expected from theoretical models [34]. Their experimental results surprisingly demonstrate that in the case of Cy5 dyes the luminescence quenching is almost exclusively due to a phase-induced suppression of the radiative rate while the energy transfer is of minor importance. The same authors had already investigated these rates in systems where lissamine dye molecules were kept at a fixed distance of 1 nm from the surface of gold nanoparticles of different sizes (average diameter from 2 to 60 nm) [35]. They showed that the quenching efficiency is already 99.8% for the 1 nm radius gold nanoparticles and, more interestingly, it is again mainly induced by a reduction of the radiative rate.

Other studies have concerned the core size effects on the photophysical properties of appended chromophores [36, 37] (also for different metal cores including alloys [36]). They all agree that, for gold nanoclusters capped with thiol-derivatized emissive molecules, the efficiency of their electronic interaction does not depend on the size of the metal core, as long as it spans over a few nanometers. Interestingly, this same electronic strong and direct interaction was investigated in a recent paper [38] from a different point of view. The authors have evidenced the preservation of absorption light polarization through the

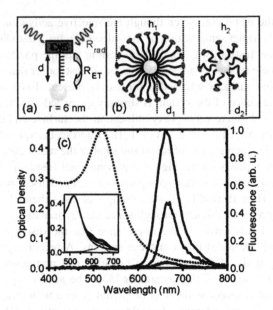

FIGURE 5.3. (a) Sketch of a Cy5 dye molecule attached via a thiol-functionalized ssDNA to a gold nanoparticle of radius $r = 6$ nm. (b) Due to limited space and repulsion, the ssDNA strands are fully stretched at a maximal surface load, providing a Cy5-AuNP distance d_1. At a reduced surface load the ssDNA is less stretched, providing a Cy5-AuNP distance $d_2 < d_1$. This allows fine-tuning the distance d. (c) Dotted line: optical density spectrum of a 1.94 nM AuNP solution. Solid lines in order of decreasing signal: fluorescence spectra of a 0.24 M aqueous solution of pure Cy5-DNA$_{43}$ molecules, Cy5-DNA$_{43}$-AuNP, and Cy5-DNA$_8$-AuNP, both with highest possible surface coverage on AuNPs. The fluorescence is reduced by 50% for Cy5-DNA$_{43}$ and by 95% for Cy5-DNA$_8$. (Inset) Solid lines: OD for different numbers of Cy5-DNA$_8$ molecules per AuNP (0, 64, 98, and 123 Cy5 per gold nanoparticle in order of increasing OD at 650 nm). Dotted line: OD of a 0.25 μM Cy5 solution. (Reproduced from Ref. 33. Copyright 2005 American Chemical Society.)

energy transfer between the dye and the gold core, for a system presenting terthiophene-based chromophores binding gold nanoparticles dispersed in a stretched polymer matrix.

It is worth underlining here that it is possible to influence the different excited-state deactivation pathways in these nanosystems changing the appended chromophore. This means that each single system has its own specific properties arising from its components and their peculiar interactions in the assembly. Therefore, it is not possible to spot general behaviors but, at the same time, this is at the base of their versatility, one of the most valuable characteristics of nanoparticles.

Among available dyes, the most extensively used luminophore as gold cluster capping agent is pyrene. This fluorescent moiety is interesting because it presents a characteristic excimer luminescence band, different from the monomer one, that can be observed only when two or more moieties come in close contact. Moreover, as opposed to the mentioned chromophores, it has been experimentally shown that electron transfer is the most important deactivation path when pyrene is bound on the surface of gold clusters [39]. Kamat and Thomas have been extensively investigating a variety of pyrene-functionalized gold nanoparticles for a few years, and recently reported for the first time a spectroscopic demonstration of the direct electron transfer process between the bounded chromophore and the metal cluster. Transient absorption spectra recorded after pulsed laser irradiation were attributed by the authors to the formation of a pyrene radical cation by interaction of excited pyrene with the gold nanocore and its successive decay via a back electron-transfer process. As already mentioned [15], thiol-capped gold nanoparticles can undergo quantized charging and, interestingly, in this system the charge separation lasts for several microseconds. Therefore, this is the first demonstration that fluorophore–gold nanoparticle systems are suitable for use in light-harvesting applications. A further investigation of systems functionalized with thiol derivatives of pyrene, presenting different linker lengths and investigated in solvents of different polarity [40], confirmed that the electron transfer to the gold nanocore is an important competitive deactivation path, which dominates in polar solvents over fluorescence emission and intermolecular excimer formation. All of these findings have been the basis for the authors to design and investigate a more complex system in order to demonstrate the possibility of using this kind of metal–chromophore nanoassemblies for applications in the sensor and display fields [41]. The co-functionalization of gold nanopaticles with both pyrene-derivatized thiols and carboxylic acid residue-derivatized thiols has yielded nanoclusters able to bind to a nanostructured TiO_2 film deposited on an optically transparent electrode. These pyrene-modified electrodes can generate a sensitized photocurrent when they undergo a proper photoexcitation but more interestingly they also demonstrate the possibility of electrochemically modulating the fluorescence of the dye. The electron transfer from the pyrene to the gold core, which is responsible for the quenching of its luminescence, is prevented when the electrode is taken to negative potentials and therefore the gold clusters are negatively charged. This induces an increasing recovery of the pyrene fluorescence up to potentials around -1.2 V, when more then 90% of the quenched emission is restored.

Other papers have illustrated the photocurrent obtained with electrodes derivatized with chromophore-functionalized gold nanoparticles [42, 43]. S. Yamada and co-workers [42] have studied a gold electrode with gold nanoparticles funtionalized with tris(2,2'-bipyridine)ruthenium(II)-viologen-thiol moieties deposited on it, but no photophysical characterization of the nanoparticles in solution or of the electrode was performed. On the contrary, this has been done by Kamat and co-workers [43] for a system obtained electrophoretically depositing gold nanoparticles functionalized with thiol derivatives of fullerene on a nanostructured SnO_2 film adhering on an optically transparent electrode. The authors report a total quenching of the fullerene emission at 710 nm occurring mainly via energy transfer in toluene solutions and they suggest the possible future exploitation of similar aggregates as photoactive antenna systems. Other researchers have investigated fullerene-functionalized gold nanoparticles [44] and their spectroscopic and electrochemical investigations [45] are in agreement with these results.

Pyrene derivatives are the functionalizing chromophores of choice for gold surfaces and nanoparticles not only for their peculiar photophysical properties but also to take advantage of their ability to give $\pi-\pi$ stacking interactions. A pyrene functionalized with a long aliphatic chain ending with a thiol has been used, for example, to self-assemble on the surface of carbon nanotubes. The pyrene moieties lie on the walls of the nanotubes exposing the thiol edges toward the solution and this allows an efficient further binding of gold nanoparticles [46]. The luminescent emission of the dye in the final assembly is almost totally quenched, while the Raman spectrum signals of carbon nanotubes is enhanced. Interestingly, this is an easy and efficient way to obtain a dense coating of carbon nanotubes with gold nanoparticles. In another study based on the same aromatic stacking properties of pyrene, together with hydrogen-bonding elements, Rotello and co-worker [47] have proved the influence of the radial structure of monolayer-protected gold nanoparticles in the multivalent molecular recognition of guest molecules such as flavine. However, no photophysical characterization of the system was presented. On the contrary, many other research groups have taken advantage of the pyrene fluorescence sensitiveness to the degree of isolation indicated by the ratio of its monomer and excimer emission to exploit it as a spectroscopic probe to reveal useful information on dynamic or catalytic processes taking place on the surface of the nanoparticle. In this context, this kind of probe was used to monitor the binding of thioester and thiocarbonate to the surface of gold clusters [48]. Pyrene derivatives of these groups were exposed to gold nanoparticle surfaces and the results showed that not only do they bind noncovalently on the metal surface, but also that gold has a very important role in catalyzing the hydrolysis of these protected thiols. The passivation of the metal core is evidenced by a large increase of the typical excimer dye emission due to the dense packing of the aromatic units and their consequent proximity and facing. The hydrolysis reaction causes the release of the pyrene moiety in solution as alcohol or carboxylic derivative, yielding the recovery of the monomeric emission and the suppression of the excimeric one. The same process was then used to monitor the binding of a mercaptosilane derivative on gold nanoparticles, the following solgel hydrolysis yielding gold core–silica shell nanoparticles [49].

Following the same idea of exploiting this ability of pyrene moieties to signal their vicinity to each other and to the nanoparticle core, our research group has investigated the possibility of a modulation of the photophysical properties of pyrene-functionalized gold nanoparticles by an accurate control of the degree of surface coverage [50]. We prepared diluted acetonitrile solutions of gold clusters stabilized by weakly bonding tetraoctylammonium bromide groups and with an average diameter of 7 nm. The gradual coverage of the nanoparticle surfaces was achieved simply by adding increasing amounts of thiol pyrene derivative to these nanoparticle dispersions. On the basis of the absorption and fluorescence data, the processes of cluster surface modification can be rationalized as follows: (1) at very low concentrations, the pyrene moieties bind to the gold colloids without interacting with each other, their fluorescence is quenched via electron transfer, and the corresponding band profile is distorted because of the already discussed interaction with the metal [39]; (2) as soon as the fluorophore density on the surface increases, interactions occur between the pyrene moieties leading to excimerlike emission (these interactions are also proved by a red shift of the absorption maximum); (3) after surface saturation, absorption and fluorescence spectra are dominated by the contribution of the signal typical of the free chromophore in solution (Figure 5.4). We demonstrated, therefore, that it is possible to control the degree of derivatization of the surface of gold nanoclusters with a fluorescent moiety and in such a way to modulate the fluorescence properties of the resulting assembly simply by changing the surface loading degree. This offers a very simple and versatile approach to new functional nanodevices.

An analogous probe system, namely, a symmetric alkyl disulfide dipyrene moiety, has been recently used to investigate the binding of disulfides on the surface of gold nanoparticles and their following dynamics on the surface [51]. The excimer/monomer emission intensity ratio of the disulfide probe was found to be very different in solution versus after the binding on the surface. The authors suggest that all of the experimental evidence can be explained by the cleavage of the disulfide bonding and the subsequent migration of the two separated pyrene units on the gold surface. It has to be noted that the real nature of the thiol and disulfide bonding on metals is still not well understood and many are the studies at present that try to give a deeper insight on this topic.

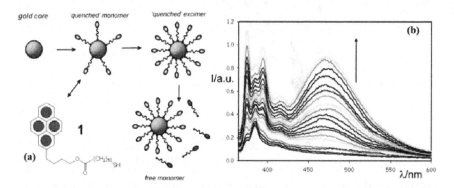

FIGURE 5.4. (a) Simplified schematic representation of the organization of the fluorescent thiols on the nanoparticle surfaces. (b) Changes in the fluorescence spectra ($\lambda_{exc} = 341$ nm) of a solution containing gold nanoparticles (8×10^{-9} M) for addition of **1**. Each curve corresponds to a concentration increase of 2.5×10^{-7} M.

In this context, we have investigated the dynamics of place-exchange reactions on monolayer-protected gold clusters via fluorescence spectroscopy [52]. Exchange of thiols at the metal nanoparticle surface is a straightforward and widely exploited way to introduce novel functionalities on the clusters, without introducing drastic alterations of the experimental synthetic conditions [23]. The dynamic of the same process has already been investigated by Murray and co-workers, but via ^{1}H-NMR studies [23b]. The idea we pursued was to detect via fluorescence spectroscopy the release of the pyrene thiol derivative from the gold cluster surface upon exchange with decanethiol. The chromophore derivatives stabilizing gold nanoparticles with an average diameter of 4 nm, as expected, were strongly quenched by the interaction with the metal. We added different amounts of decanethiol to four samples of the same dichloromethane dispersion of the chromophore-derivatized nanoparticles and we observed in all cases a slow recovery of pyrene fluorescence, as a result of the exchange reaction (Figure 5.5). This can be explained keeping in mind that pyrene moieties, whose fluorescence is quenched when bound to the metal, after desorption no longer interact with the gold clusters, so that fluorescence is completely recovered. It was observed that this recovery in time depends on the concentration of the added nonfluorescent

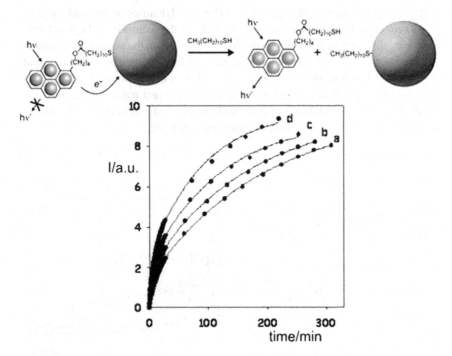

FIGURE 5.5. (Upper) Schematic representation of the release of pyrene derivative upon exchange with decanethiol. (Lower) Changes in the fluorescence intensity at 395 nm ($\lambda_{exc} = 328$ nm) of a solution of gold nanoparticles capped with thiolated pyrene moieties for different concentrations of decanethiol: (a) 1.7×10^{-4} M; (b) 3.3×10^{-4} M; (c) 6.6×10^{-4} M; (d) 1.3×10^{-3} M.

thiol, as would be expected for an associative mechanism, and that the complete exchange presented two different pseudo-first-order processes at the beginning and toward the end of the monitored period. We calculated the two associated kinetic constants and, as a consequence, the reaction orders with respect to the incoming thiol for the two kinetic processes that were evidenced. In the two cases an order of reaction of 0.33 and 0.38 was found, suggesting that in the rate-determining step the insertion of one molecule of decanethiol weakens the interaction of more than a single fluorophore bound unit. Molecular modeling results agreed with the kinetic ones: both analyses show the presence of two regimes and suggest that for each decanethiol that bonds on the surface more than one of the pyrene-labeled thiols is involved in the rate-determining step. The results of our experiments are therefore consistent with an associative exchange path, as already reported [23b], but they also indicate that the insertion of the approaching thiol in the preformed lattice weakens the interaction of several adsorbed molecules (the calculation results indicated at least three of them), which results in the rearrangement of more than a single unit in the reaction rate-determining step (Figure 5.6). This suggests that monolayer-protected gold nanoclusters are even more versatile than previously thought.

This versatility is also the basis of the relative ease in inducing array formation in this kind of system. When bifunctionalized capping agents are used, for example bearing two thiol groups on the opposite endings of the capping moiety, they can be used either to fix the resulting derivatized gold nanoparticles onto metal (gold, platinum, or silver) surfaces (e.g., electrodes [53]) or, very interestingly, to act as bridging ligands to form nanoparticle arrays [54].

Very recently, this possibility was exploited by Liu and co-workers to obtain supramolecular aggregates based on polypseudorotaxane-functionalized gold nanoparticles able to act as captors for fullerenes and also exhibiting a good

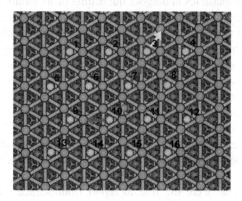

FIGURE 5.6. Au slab (yellow) with the sulfur atoms of the thiol chains: in green, the sulfur of the pyrene-functionalized thiols; in purple, the location of sulfur of the intruding decanethiol. The arrows indicate the rearrangements. For clarity, the long alkyl chains are not shown. (Reproduced from Ref. 52. Copyright 2003 American Chemical Society.)

DNA cleavage ability upon light irradiation [55]. The authors synthesized polypseudorotaxanes inducing, in polar solvents (water), the multiple threading of L-tryptophan-modified β-cyclodextrins onto polypropylene glycols terminating with amino groups on both endings, and subsequently they let them react with a dispersion of citrate-stabilized gold nanoparticles. They recorded much experimental evidence of the formation of cluster aggregates with the polypseudorotaxanes acting as bridging ligands between the different gold cores. This process caused a decrease in the emission intensity of the L-tryptophan moieties via the interaction with Au nanoparticles. This partial quenching effect, as already discussed, is distance dependent, and since the luminescent units are distributed along the whole polymers they experience different distances from the gold surfaces. A more remarkably quenched emission was revealed for the same system after addition of [60] fullerene suggesting an electron transfer process occurring between the tryptophan donor units and the C_{60} acceptor ones. The special ability of the aggregates functionalized with L-tryptophan units to capture fullerene could be explained by the relatively strong interaction between C_{60} or its derivatives and amino acids that has been observed and already reported in biological cells. Due to the good water solubility of the L-tryptophan-modified polypseudorotaxane–gold nanoparticle aggregates they proved that this is an efficient system to enrich [60]fullerene in aqueous solutions. Moreover, the same array thus formed exhibited a good DNA cleavage ability under light irradiation. The poor water solubility of [60]fullerene is very well known, and overcoming this problem has proved to be difficult.

5.2.3 Nonquenched Fluorophores Directly Linked to Nanoparticles

So far, we have discussed heterosystems in which a flurophore binding on the surface of gold nanoparticles undergoes quenching of its luminescence intensity, caused by the strong interaction between the dye and the metal surface. A large amount of research work has been dedicated in the last two decades to elucidation of the nature of this interaction and the factors that can alter or influence it [13c, 33, and references herein] . All of these results, yielding a much deeper understanding of the quenching processes and of the connected parameters, have inspired some authors to design systems able to use the metal-to-chromophore vicinity to obtain, on the contrary, an enhancement of the luminescence intensity or at least its preservation.

For example, an increase of the radiative constant of different organic chromophores, when they lie in close proximity to metal surfaces, has been observed [56].

Thomas and Kamat were the first to report fluorescence enhancement of a chromophore directly linked to the surface of a gold nanoparticle and very close to it [57]. 1-Methylaminopyrene was attached on gold clusters (with a diameter ranging from 5 to 8 nm), weakly stabilized by tetraoctylammonium bromide (TOAB) moieties, via exchange reaction in THF. The yielding system features the presence of both bounded chromophores and TOAB moieties on the gold surfaces; in this environment the pyrene moieties are quite isolated from one another, as demonstrated by the absence of the excimer emission. On the contrary, an intense and red-shifted monomeric emission was detected, while the nanoparticles are nonfluorescent and the 1-methylaminopyrene in THF was weakly luminescent. Absorption, steady-state emission, and lifetime measurements showed that this can be explained by taking into account the interaction of the amino lone pair with the metal, when the pyrene is bound on its surface. This induces a decrease in its donating ability and suppresses the photoinduced electron transfer responsible for the main nonradiative decay process of the luminophore singlet excited state. These amine binding pyrenes, organized on gold nanoparticles, represent the first example of organic–inorganic materials of this kind that are fluorescent and open up new potentialities for these heteromaterials to be used for useful applications in photodevices or biophotonics.

Unfortunately, not many other examples are available in the literature. An interesting paper was published reporting the first observation of the fluorescence lifetime enhancement for a chromophore very close and strongly bound to the surface of a gold nanoparticle [58]. 4-Acetamido-4'-maleimidylstilbene-2,2'-dithiol (AMDT)-functionalized gold nanoparticles (average diameter 3.5 nm) were synthesized following the method proposed by Brust and Schiffrin. AMDT is a dithiol that binds to the gold surface in a bidentate mode ensuring the molecule is positioned in such a way that its molecular dipole moment is parallel to the surface (Figure 5.7). The authors present a very interesting and comprehensive set of results, including steady-state and time-resolved spectroscopy at different temperatures, and conclude that a molecular dipole moment oriented in such a way is reduced. This would enhance the fluorescence lifetime of the excited state of the chromophore (that doubled for the system in study at room temperature) when it is very close to the noble metal nanoparticle surface. No decrease of the fluorescence quantum yield could be measured and this indicates a weak energy transfer between the metal and the dye in the organic–inorganic system.

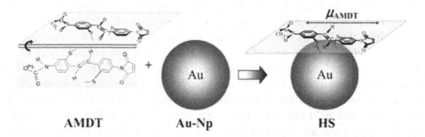

FIGURE 5.7. Structure of the 4-acetamido-4'-maleimidylstilbene-2,2'-dithiol (AMDT), schematic representation of the 3.5-nm-diameter gold spherical nanoparticles (Au-Np), and biattached AMDT onto a gold nanoparticle hybrid system (HS). The double-headed arrow represents the plane of the molecular dipole moment of AMDT. (Reproduced from Ref. 58. Copyright 2005 American Chemical Society.)

5.2.4 Fluorophores Unspecifically Interacting
with Nanoparticles

Here we discuss chromophores that interact with gold nanoparticles simply via mixing of the two components in solution. This implies that the interaction can be due to diffusional collisions and/or aspecific adsorption events. In general, the surface adsorption is based on electrostatic forces and yields assemblies whose stability strongly depends on the polarity of the solvent and, sometimes, on pH conditions. It is not surprising that the first dyes that have been studied are fluorescein and rhodamine, the two most widely used probes in medical and biological assays. In an interesting paper [59] published in 1999, the authors employ the absorption at pH 9 of fluorescein isothiocyanate (FITC) on gold nanoparticle surfaces to encapsulate this dye inside silica shells, but they do not present a deep investigation of the metal–organic system since it is only an intermediate to reach their final target. On the contrary, Kamat and co-workers have rigorously studied a heterosystem obtained absorbing the positively charged rhodamine 6G onto gold nanoclusters stabilized by isothiocyanate moieties [60]. The emission intensity of the chromophore at the metal surface is quenched and, very interestingly, the rhodamine units display an H-type interaction that induces strong aggregation of many clusters. The electrostatic absorption of the dye at the surface provokes both the aggregation of the particles and their charge neutralization and these two factors yield cluster coalescence to form larger nanoparticles. This has been observed via changes both in the absorption spectra (plasmon resonance band) and in the fluorescence ones (release and consequent enhancement of rhodamine 6G emission intensity).

More recently, charged metal complexes have been employed to obtain heteroassemblies and to investigate their photophysical properties. The same research group has reported evidence of the very strong bonding of chlorophyll *a* on the surface of gold nanoparticles with an average diameter of 8 nm [61]. The association mechanism is not yet clear but plausibly the electron-rich gold clusters can interact with the coordinatively unsaturated central magnesium atom in the chlorophyll. Interesting photophysical studies have evidenced that the fluorescence quenching caused by the interaction with the metal surface is mostly attributable to static quenching while dynamic quenching plays a minor role. The large value estimated for the association constant and the favorable redox potentials of the chlorophyll *a* and of the metal nanoparticles provide the necessary conditions for an electron transfer from the organic moieties to the gold core. The authors have investigated, therefore, the effects that charging the gold nanocore has on the chlorophyll fluorescence. They have deposited the chromophores on a nanostructured gold electrode applying a negative potential, in analogy with another system already reported [41]. The resulting negative charging of the nanoparticles induced a total recovery of the fluorescence preventing the electron transfer from the chromophores and therefore their quenching.

A clearer understanding of the interaction type between the metal surface and a transition metal complex has been reported by Murray and co-worker [62]. The positively charged complex [Ru(bpy)$_3$]$^{2+}$ is more efficiently quenched by a solution containing gold nanoparticles presenting negative charges on their surface in comparison with positively charged ones, since in the latter case only a dynamic quenching can occur. Even more interestingly, the addition of cations to the first system competes with the electrostatic interaction occurring on the surface of the clusters displacing the chromophore and allowing a recovery of the luminescence signal. The quenching efficiency shows a slight dependence on the dimensions of the gold core. The authors have shown a nice method to influence reversible binding interactions of chromophore on the surface of gold nanoparticles and to measure them.

5.3 Photoluminescent Gold Nanoparticles

Photoluminescence from gold nanoclusters was to some degree predicted by Mooradian [63] in 1969, who observed and investigated the light emission from noble metal surfaces. The role of roughness in the luminescence efficiency was then studied by Boyd [64] but the phenomenon was finally correlated to the electronic structure only in 1988 [65]. Excitation was then related to transitions from occupied d bands to states higher in energy than the Fermi level while emission was related to the recombination of the hole with an electron from an occupied sp band; it has to be noted that, between the two processes, electron–phonon and hole–phonon scattering take place causing a partial energy loss.

Photoluminescence from gold colloids stabilized with nonfluorescent moieties was reported for the first time, to our knowledge, by Wilcoxon and co-workers [66]. They prepared the metal clusters in water following the well-known method proposed by Turkevitch [21] and they reported a quantum yield around 10^{-4}–10^{-5} for the emission observed at 440 nm upon excitation at 230 nm. Employing size-exclusion liquid chromatographic fractionation, the size-dependency of the luminescence could be investigated showing that no luminescence was detected for clusters exceeding 5 nm. This clearly suggests that quantum effects [67] are of fundamental importance in the generation of the photoluminescence. A mechanism similar to the one proposed for roughened gold surfaces was suggested, ascribing the photon emission to the recombination of the holes generated after the excitation in the d or sp bands with Fermi-level electrons.

A different kind of transition, on the other hand, is suggested by Whetten [68] to be responsible for the near-infrared emission of small gold nanocrystals (1.1 and 1.7 nm, around 38 and 145 gold atoms, respectively). Upon excitation at 1064 nm, the nanocrystals show a broad luminescence band in the region between 1100 and 1600 nm. Because of the low energy involved, this emission was related to an intra-sp-band transition. The quantum yield was reported only for the 1.7-nm nanoclusters and it was estimated to be 4.4×10^{-5} with an uncertainty of about 30%.

A much larger luminescence efficiency was reported by El-Sayed and co-workers in the case of gold nanorods (quantum yield 1×10^{-4}–1×10^{-3}) [69]. In this case the luminescence band maximum is strongly dependent on the average ratio between the length and the width of the nanorods (aspect ratio) and it shifts, for example, from 548 nm to 588 nm changing it from 2 to 5.4. For the correlation, rods with similar width (20 nm) were compared. It is surprising that, in contrast to that observed for "spherical" nanoparticles, in the case of nanorods luminescence quantum yield increases with the square of the length of the rods.

Nevertheless, an even higher luminescence efficiency ($\Phi = 0.003 \pm 0.001$) was reported by Huang and Murray for water-soluble nanoclusters (1.8-nm diameter) protected with tiopronin thiolate [70]. The authors observed that luminescence was strengthened after purification by dialysis and they suggested that some unidentified synthetic by-products could behave as quenchers. Moreover, they investigated the effect of different thiolate ligands observing that the luminescence band energy and intensity was affected by the nature of the protecting agent. This led the authors to conclude that the 6sp emitter states "might be regarded as a kind of surface electronic states" of the gold nanoclusters.

All of the systems discussed so far, even if apparently similar, present very different photophysical behaviors and the various authors explain them suggesting different transitions. Interestingly, El-Sayed and Whetten proposed a schematic model of excited state levels that combines both intraband (sp) and interband (sp–d) transitions to explain the two-component emission shown by the system that they were investigating [71]. Twenty-eight-atom gold clusters having 16 glutathione molecules absorbed on the surface ($Au_{28}SG_{16}$) present a structured luminescence spectrum which can be deconvoluted into two bands with maxima at around 800 nm (1.55 eV) and 1100 nm (1.13 eV). In the molecular-like model proposed (Figure 5.8), the excitation is represented as a singlet–singlet transition that moves an electron from the d band to an unoccupied orbital of the sp band. Relaxation of the system can lead to the electron–hole recombination giving a "fast" (fluorescence-like) light emission. Alternatively, a nonradiative recombination of as sp electron with the hole in the d band can generate a triplet-like state that could deactivate "slowly" giving a phosphorescence-like emission. This model, even though nonrigorous, has the advantage of giving a simple interpretation of the results obtained with luminescence time-resolved experiments. The luminescence decays recorded at different wavelengths show that at least two exponential components must be taken into account: a "fast" one with a lifetime on the order of 100 ns and a "slow" one of about 1000 ns. Moreover the contribution of the "slow" decay is larger at longer wavelengths in agreement with the proposed model. The molecular-like behavior of $Au_{28}SG_{16}$ was confirmed by femtosecond transient absorption studies [72].

A similar molecular nature was reported by Murray for slightly bigger systems, 38-atom gold nanoparticles [73]. For the luminescence of this kind of clusters, the nature of the thiolate ligands proved to be of extraordinary relevance. In the case of 2-phenylethanethiol (PhC_2SH)-protected nanoparticles, indicated as $Au_{38}(PhC_2S)_{24}$, almost no luminescence could be detected, whereas after ligand place exchange reaction with thiolated poly(ethyleneglycol) ($PEG_{135}SH$) the resulting $Au_{38}(PEG_{135}S)_{13}(PhC_2S)_{11}$ particles show a broad emission band around 1000 nm. Also in this case, the spectrum was fitted with two Gaussian-shaped

(a)

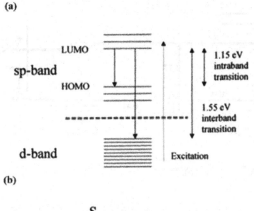

(b)

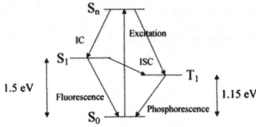

FIGURE 5.8. (a) Solid-state model for the origin of the two luminescence bands: the high-energy band is proposed to be due to radiative interband recombination between the sp and d bands while the low-energy band is thought to originate from radiative intraband transitions within the sp band across the HOMO–LUMO gap. Note that intraband recombination has to involve prior nonradiative recombination of the hole in the d band created after excitation with an (unexcited) electron in the sp band. (b) Molecular model for the origin of the two luminescence bands: Excitation into higher excited states (S_n) is followed by rapid relaxation (internal conversion, IC) to the lowest excited singlet state (S_1). Radiative recombination with the ground state (S_0), fluorescence, gives rise to the high-energy luminescence band. Intersystem crossing (ISC) to the lowest excited triplet state (T_1) followed by radiative relaxation to the ground state, phosphorescence, causes the luminescence band at lower energies. (Reproduced from Ref. 71. Copyright 2002 American Chemical Society.)

emission bands centered at 902 nm (1.38 eV) and 1025 nm (1.2 eV). On the basis of the electronic properties of the system, resulting from electro-chemical and optical characterizations, the authors conclude that while the 1.38-eV transition is compatible with the band gap the 1.2-eV one is to be considered a sub-band-gap energy luminescence (Figure 5.9).

Recently, Murray published a study on the comparison of the spectra of different clusters ranging from Au_{11} to Au_{201} [74]. An interesting result of this analysis is that diverse core sizes show luminescence in the same spectral region even though their absorption spectra are very different (Figure 5.10). This means that the energy of the transitions involved in the emission process is not clearly

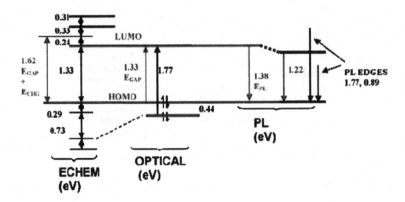

FIGURE 5.9. Schematic model energy level diagram for $Au_{38}(PhC_2S)_{24}$, based on data taken in CH_2Cl_2 solution (except for reduction beyond Au_{38}^- where data are from 2:1 toluene/acetonitrile). (Reproduced from Ref. 73. Copyright 2004 American Chemical Society.)

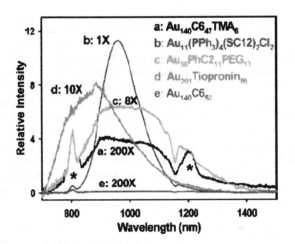

FIGURE 5.10. AuNP with different core sizes and monolayers. C6, C12, PhC2, PEG, and PPh$_3$ represent hexanethiolate, dodecanethiolate, phenylethanethiol, poly(ethylene glycol) (MW 350) thiolate, and triphenylphosphine, respectively. The spectrum of tiopronin NP was measured in D_2O; the others were measured in methylene chloride. All spectra were excited at 400 nm. The quantum efficiencies (relative to Q-switch 5 and DTTC[11]) of the Au_{11}, Au_{38}, Au_{140} TMA, and Au_{201} NP are 0.08, 1.2 × 10^{-2}, 1.8 × 10^{-4}, and 9 × 10^{-3}, respectively. Asterisks indicate artifacts from second- and third-order excitation peaks (800 and 1200 nm); the dip at 1165 nm is partly due to solvent/ligand absorption. (Reproduced from Ref. 74. Copyright 2005 American Chemical Society.)

related to the HOMO–LUMO gap but, according to the authors, "reflects a participation of localized core surface state that have size dependent energetics." Experiments into surface modification confirm that the near-infrared emission of gold nanoparticles is an electron surface-state phenomenon.

Due to the low absorption and scattering of near-infrared (NIR) light by biological tissues, the interest in NIR emitters is especially high. Experiments are currently in progress worldwide to obtain nanoparticles having a more intense NIR luminescence. This research could lead to luminescent labels featuring very interesting photophysical properties to be used in nanobiotechnological applications in particular in medical and biological assays.

5.4 Conclusions

In this chapter we have given an overview of the state of the art in the field of fluorophore derivatized and luminescent gold nanoparticles. This research topic is part of a much wider one that goes through metal nanoparticles to quantum dots in general as materials for nanotechnology. Even though the applications of these new and growing technologies are still limited, nanoparticles have found a wide use in many different industries and laboratories.

The association of luminescence with metal nanoclusters, however, has been considered until very recently something interesting but also difficult to achieve. This wariness arises from the observation that the metal core is generally able to quench very efficiently the luminescence of most emitting dyes.

We hope to have shown that the intense research on this subject is leading to the design and synthesis of many different nanoparticles with their own and intriguing photophysical properties. We believe that the initial concern will be overcome, leading to the awareness that the combination of luminescence and metal nanoclusters is not only possible but also particularly appealing for applications in many fields of large impact.

Acknowledgments. Financial support from MIUR (SAIA and LATEMAR projects) and University of Bologna is gratefully acknowledged.

References

1. *Nanotechnology: A Realistic Market Evaluation*, Business Communication Co. Inc., Norwalk (2004).
2. *Nanotechnology Market Opportunities, Market Forecasts, and Market Strategies*, 2004–2009, Winter Green Research, Lexington (2004).
3. Crichton, M., *Prey.* HarperCollins, New York (2002).
4. (a) Storrs, J.H., *Nanofuture: What's Next for Nanotechnology*, Prometheus Books, New York (2005). (b) Schulte, J., *Nanotechnology: Global Strategies, Industry Trends and Applications*, Wiley, New York (2005). (c) Theodore, L., and

Kunz, R.G., *Nanotechnology: Environmental Implications and Solutions*, Wiley–Interscience, New York (2005). (d) Yao, N., and Wang, Z.L., *Handbook of Microscopy for Nanotechnology*, Springer, New York (2005). (e) Mansoori, G.A., *Principles of Nanotechnology: Molecular-Based Study of Condensed Matter in Small Systems*, World Scientific, Singapore (2005). (f) Malsch, N.H., *Biomedical Nanotechnology*, CRC Press, London (2005). (g) Waite, S.R., *Quantum Investing: Quantum Physics, Nanotechnology, and the Future of the Stock Market*, Texere Publishing, Mason, Ohio (2004). (h) Lakhtakia, A., *Handbook of Nanotechnology: Nanometer Structure Theory, Modeling, and Simulation*, Wiley, New York (2004). (i) Wolf, E.L., *Nanophysics and Nanotechnology: An Introduction to Modern Concepts in Nanoscience*, Wiley, New York (2004).

5. (a) Tsunoyama, H., Sakurai, H., Negishi, Y., and Tsukuda, T., Size-specific catalytic activity of polymer-stabilized gold nanoclusters for aerobic alcohol oxidation in water, *J. Am. Chem. Soc.* **127**, 9374–9375 (2005). (b) Kisailus, D., Najarian, M., Weaver, J.C., and Morse, D.E., Functionalized gold nanoparticles mimic catalytic activity of a polysiloxane-synthesizing enzyme, *Adv. Mater.* **17**, 1234–1239 (2005). (c) Esparza, R., Ascencio, J.A., Rosas, G., Sànchez Ramìrez, J.F., Pal, U., and Perez, R., Structure, stability and catalytic activity of chemically synthesized Pt, Au, and Au-Pt nanoparticles, *J. Nanosci. Nanotechnol.* **5**, 641–647 (2005). (d) Pasquato, L., Pengo, P., and Scrimin, P., Functional gold nanoparticles for recognition and catalysis, *J. Mater. Chem.* **14**, 3481–3487 (2004). (e) Campbell, C.T., Physics: The active site in nanoparticle gold catalysis, *Science* **306**, 234–235 (2004). (f) Meyer, R., Lemire, C., Shaikhutdinov, K.S., and Freund, H.J., Surface chemistry of catalysis by gold, *Gold Bull.* **37**, 72–124 (2004). (g) Haruta, M., Gold as a novel catalyst in the 21st century: Preparation, working mechanism and applications, *Gold Bull.* **37**, 27–36 (2004). (h) Haruta, M., Catalysis by gold nanoparticles, in *Encyclopedia of Nanoscience and Nanotechnology*, Vol. 1, edited by H.S. Nalwa, American Scientific Publishers, Stevenson Ranch, California, pp. 655–664 (2004). (i) Vayenas, C.G., Wieckowski, A., and Savinova, E.R., *Catalysis and Electrocatalysis at Nanoparticle Surfaces*, Dekker, New York (2003).

6. (a) Giaever, I., and Zeller, H.R., Superconductivity of small tin particles measured by tunneling, *Phys. Rev. Lett.* **20**, 1504–1507, (1968). (b) Kubo, R., Electronic properties of metallic fine particles I, *J. Phys. Soc. Japan* **17**, 975–986 (1962). (c) Kubo, R., Electronic properties of metallic fine particles, *Phys. Lett.* **1**, 49–50 (1962). (d) Fröhlich, H., The specific heat of small metallic particles at low temperatures, *Physica* **4**, 406–412 (1937).

7. (a) Shaw, C.F., III, Gold-based medicinal agents, *Chem. Rev.* **99**, 2589–2600 (1999). (b) Higby, G.J., Gold in medicine: A review of its use in the West before 1900, *Gold Bull.* **15**, 130–140 (1982). (c) Brown, D.H., and Smith, W.E., The chemistry of the gold drugs used in the treatment of rheumatoid arthritis, *Chem. Soc. Rev.* **9**, 217–240 (1980). (d) Gibson, C.S., Gold in medicine, *Chem. Prod. Chem. News* **1**, 35–36 (1938).

8. (a) Sato, K., Hosokawa, K., and Maeda, M. Non-cross-linking gold nanoparticle aggregation as a detection method for single-base substitutions, *Nucleic Acids Res.* **33**, e4/1-e4/5 (2005). (b) Sonvico, F., Dubernet, C., Colombo, P., and Couvreur, P., Metallic colloid nanotechnology, applications in diagnosis and therapeutics, *Curr. Pharm. Des.* **11**, 2091–2105 (2005). (c) Hyatt, A.D., and Eaton, B.T., *Immuno-Gold Electron Microscopy in Virus Diagnosis and Research*, CRC Press, Boca Raton (1993). (d) Eagle, H., Applications of colloid chemistry in the serum diagnosis

of syphilis, *J. Phys. Chem.* **36**, 259–267 (1932). (e) Kahn, R.L., Serum diagnosis for syphilis, in *Colloid Chemistry: Theoretical and Applied*, Vol. II, edited by J. Alexander, The Chemical Catalog Co., New York (1929).

9. Helcher, H.H., (1718). *Aurum Potabile oder Gold Tinstur*, Johann Herbord Klossen, Breslau (1718).

10. Faraday, M., Experimental relations of gold (and other metals) to light, *Philos. Trans.* **147**, 145–181 (1857).

11. Brust, M., Walker, M., Bethell, D., Schiffrin, D.J., and Whyman, R.J., Synthesis of thiol-derivatized gold nanoparticles in a two-phase liquid–liquid system, *J. Chem. Soc. Chem. Commun.* 801–802 (1994).

12. (a) Fiorani, D., *Surface Effects in Magnetic Nanoparticles*, Springer, New York (2005). (b) Schmid, G., *Nanoparticles: From Theory to Application*, Wiley, New York (2004). (c) Rotello, V., *Nanoparticles: Building Blocks for Nanotechnology*, Springer, New York (2003). (d) Heilmann, A., *Polymer Films with Embedded Metal Nanoparticles*, Springer, New York (2002). (e) Feldheim, D.L., and Foss, C.A., Jr., *Metal Nanoparticles: Synthesis, Characterization and Applications*, Dekker, New York (2002). (f) Fendler, J.H., *Nanoparticles and Nanostructured Films: Preparation, Characterization and Applications*, Wiley-VCH, Berlin (1998).

13. (a) Daniel, M.C., and Astruc, D., Gold nanoparticles: Assembly, supramolecular chemistry, quantum-size-related properties, and applications toward biology, catalysis, and nanotechnology, *Chem. Rev.* **104**, 293–346 (2004). (b) Aslan, K., Zhang, J., Lakowicz, J.R., and Geddes, C.D., Saccharide sensing using gold and silver nanoparticles A review, *J. Fluores.* **14**, 391–400 (2004). (c) Thomas, K.G., and Kamat, P.V. Chromophore-functionalized gold nanoparticles, *Acc. Chem. Res.* **36**, 888–898 (2003). (d) Schmid, G., and Corain, B., Nanoparticulated gold: Syntheses, structures, electronics, and reactivities, *Eur. J. Inorg. Chem.* **17**, 3081–3098 (2003). (e) Zhong, Z., Male, K.B., and Luong, J.H.T., More recent progress in the preparation of Au nanostructures, properties, and applications, *Anal. Lett.* **36**, 3097–3118 (2003). (f) Jain, K.K., Nanodiagnostics: Application of nanotechnology in molecular diagnostics, *Expert Rev. Mol. Diagn.* **3**, 153–161 (2003). (g) McMillan, R.A., and Andrew, R., Biomolecular templates: Nanoparticles align, *Nature Mater.* **2**, 214–215 (2003). (h) Csaki, A., Moller, R., and Fritzsche, W., Gold nanoparticles as novel label for DNA diagnostics, *Expert Rev. Mol. Diagn.* **2**, 187–193 (2002). (i) Brust, M., and Kiely, C.J., Some recent advances in nanostructure preparation from gold and silver particles: A short topical review, *Colloids Surf. A* **202**, 175–186 (2002).

14. Mie, G., Contributions to the optics of turbid media, especially colloidal metal solutions, *Ann. Phys.* **25**, 377–445 (1908).

15. (a) Devarajan, S., and Sampath, S., Electrochemistry with nanoparticles, in *Chemistry of Nanomaterials*, edited by C.N.R. Rao, A. Mueller, and A.K. Cheetham, Wiley-VCH, Weinheim, pp. 646–687 (2004). (b) Liljeroth, P., Quinn, B.M., Ruiz, V., and Kontturi, K., Charge injection and lateral conductivity in monolayers of metallic nanoparticles, *Chem. Commun.* 1570–1571 (2003). (c) Hicks, J.F., Miles, D.T., and Murray, R.W., Quantized double-layer charging of highly monodisperse metal nanoparticles, *J. Am. Chem. Soc.* **124**, 13322–13328 (2002). (d) Chen, S., Pei, R., Zhao, T., and Dyer, D.J., Gold nanoparticle assemblies by metal ion–pyridine complexation and their rectified quantized charging in aqueous solutions, *J. Phys. Chem. B* **106**, 1903–1908 (2002). (e) Chen, S., and Murray, R.W., Electrochemical quantized capacitance charging of surface ensembles of gold nanoparticles, *J. Phys.*

Chem. B **103**, 9996–10000 (1999). (f) Chen, S., Ingram, R.S., Hostetler, M.J., Pietron, J.J., Murray, R.W., Schaaff, T.G., Khoury, J.T., Alvarez, M.M., and Whetten, R.L., Gold nanoelectrodes of varied size: Transition to molecule-like charging, *Science* **280**, 2098–2101 (1998).

16. Quinn, B.M., Liljeroth, P., Ruiz, V., Laaksonen, T., and Kontturi, K., Electrochemical resolution of 15 oxidation states for monolayer protected gold nanoparticles, *J. Am. Chem. Soc.* **125**, 6644–6645 (2003).

17. (a) Chen, S., Murray, R.W., and Feldberg, S.W., Quantized capacitance charging of monolayer-protected Au clusters, *J. Phys. Chem. B* **102**, 9898–9907 (1998). (b) Hofstetter, W., and Zwerger, W., Single-electron box and the helicity modulus of an inverse square XY model, *Phys. Rev. Lett.* **78**, 3737–3740 (1997). (c) Hartmann, E., Marquardt, P., Ditterich, J., Radojkovic, P., and Steinberger, H., Characterisation and utilization of the context-dependent physical properties of nanoparticles for nanostructures investigated by scanning tunneling microscopy, *Appl. Surf. Sci.* **107**, 197–202 (1996). (d) Amman, M., Wilkins, R., Ben-Jacob, E., Maker, P.D., and Jaklevic, R.C., Analytic solution for the current–voltage characteristic of two mesoscopic tunnel junctions coupled in series, *Phys. Rev. B* **43**, 1146–1149 (1991).

18. (a) Speets, E.A., Dordi, B., Ravoo, B.J., Oncel, N., Hallbaeck, A.S., Zandvliet, H.J.W., Poelsema, B., Rijnders, G., Blank, D.H.A., and Reinhoudt, D.N., Noble metal nanoparticles deposited on self-assembled monolayers by pulsed laser deposition show Coulomb blockade at room temperature, *Small* **1**, 395–398 (2005). (b) Yang, Y., and Nogami, M., Room temperature single electron transistor with two-dimensional array of Au-SiO$_2$ core-shell nanoparticles, *Sci. Tech. Adv. Mater.* **6**, 71–75 (2005). (c) Chaki, N.K., Singh, P., Dharmadhikari, C.V., and Vijayamohanan, K.P., Single-electron charging features of larger, dodecanethiol-protected gold nanoclusters: Electrochemical and scanning tunneling microscopy studies, *Langmuir* **20**, 10208–10217 (2004). (d) Khomutov, G.B., Kislov, V.V., Gainutdinov, R.V., Gubin, S.P., Obydenov, A.Y., Pavlov, S.A., Sergeev-Cherenkov, A.N., Soldatov, E.S., Tolstikhina, A.L., and Trifonov, A.S., The design, fabrication and characterization of controlled-morphology nanomaterials and functional planar molecular nanocluster-based nanostructures, *Surf. Sci.* **532–535**, 287–293 (2003). (e) Chaki, N.K., Gopakumar, T.G., Maddanimath, T., Aslam, M., and Vijayamohanan, K., Effect of chain length on the tunneling conductance of gold quantum dots at room temperature, *J. Appl. Phys.* **94**, 3663–3666 (2003). (f) Suganuma, Y., Trudeau, P.E., and Dhirani, A.A., Probing correlated current and force effects of nanoparticle charge states by hybrid STM-AFM, *Phys. Rev. B* **66**, 241405/1–241405/4 (2002). (g) Rolandi, M., Scott, K., Wilson, E.G., and Meldrum, F.C., Manipulation and immobilization of alkane-coated gold nanocrystals using scanning tunneling microscopy, *J. Appl. Phys.* **89**, 1588–1595 (2001). (h) Ingram, R.S., Hostetler, M.J., Murray, R.W., Schaaff, T.G., Khoury, J., Whetten, R.L., Bigioni, T.P., Guthrie, D.K., and First, P.N., 28 kDa alkanethiolate-protected Au clusters give analogous solution electrochemistry and STM Coulomb staircases, *J. Am. Chem. Soc.* **119**, 9279–9280 (1997).

19. (a) Schmid, G., and Simon, U., Gold nanoparticles: Assembly and electrical properties in 1–3 dimensions, *Chem. Commun.* 697–710 (2005). (b) Boyen, H.G., Ethirajan, A., Kastle, G., Weigl, F., Ziemann, P., Schmid, G., Garnier, M.G., Buttner, M., and Oelhafen, P., Alloy formation of supported gold nanoparticles at their transition from clusters to solids: Does size matter? *Phys. Rev. Lett.* **94**,

016804/1-016804/4 (2005). (c) Torma, V., Vidoni, O., Simon, U., and Schmid, G., Charge-transfer mechanisms between gold clusters, *Eur. J. Inorg. Chem.* 1121–1127 (2003). (d) Boyen, H.G., Kaestle, G., Weigl, F., Koslowski, B., Dietrich, C., Ziemann, P., Spatz, J.P., Riethmueller, S., Hartmann, C., Moeller, M., Schmid, G., Garnier, M.G., and Oelhafen, P., Oxidation-resistant gold-55 clusters, *Science* **297**, 1533–1536 (2002). (e) Schmid, G., Metals, in *Nanoscale Materials in Chemistry*, edited by Wiley, K.J. Klabunde, New York, pp. 15–59 (2001). (f) Sawitowski, T., Miquel, Y., Heilmann, A., and Schmid, G., Optical properties of quasi one-dimensional chains of gold nanoparticles, *Adv. Func. Mater.* **11**, 435–440 (2001).

20. Prodi, L., Luminescent chemosensors: From molecules to nanoparticles, *New J. Chem.* **29**, 20–31 (2005).

21. Turkevitch, J., Stevenson, P.C., and Hillier, J., Nucleation and growth process in the synthesis of colloidal gold, *Discuss. Faraday Soc.* **11**, 55–75 (1951).

22. (a) Iwamoto, M., Kuroda, K., Kanzow, J., Hayashi, S., and Faupel, F., Size evolution effect of the reduction rate on the synthesis of gold nanoparticles, *Adv. Powder Technol.* **16**, 137–144 (2005). (b) Donkers, R.L., Song, Y., and Murray, R.W., Substituent effects on the exchange dynamics of ligands on 1.6 nm diameter gold nanoparticles, *Langmuir* **20**, 4703–4707 (2004). (c) Chen, S., Templeton, A.C., and Murray, R.W., Monolayer-protected cluster growth dynamics, *Langmuir* **16**, 3543–3548 (2000).

23. (a) Guo, R., Song, Y., Wang, G., and Murray, R.W., Does core size matter in the kinetics of ligand exchanges of monolayer-protected Au clusters? *J. Am. Chem. Soc.* **127**, 2752–2757 (2005). (b) Hostetler, M.J., Templeton, A.C., and Murray, R.W., Dynamics of place-exchange reactions on monolayer-protected gold cluster molecules, *Langmuir* **15**, 3782–3789 (1999). (c) Ingram, R.S., Hostetler, M.J., and Murray, R.W., Poly-hetero-ω-functionalized alkanethiolate-stabilized gold cluster compounds, *J. Am. Chem. Soc.* **119**, 9175–9178 (1997).

24. George, T.K., Zajicek, J., and Kamat, P.V., Surface binding properties of tetraocty-lammonium bromide capped gold nanoparticles, *Langmuir* **18**, 3722–3727 (2002).

25. Fitzmaurice, D., Rao, S.N., Preece, J.A., Stoddart, J.F., Wenger, S., and Zaccheroni, N., Heterosupramolecular chemistry: Programmed pseudorotaxane assembly at the surface of a nanocrystal, *Angew. Chem. Int. Ed. Engl.* **38**, 1147–1150 (1999).

26. (a) Ghosh, S.K., Nath, S., Kundu, S., Esumi, K., and Pal, T., Solvent and ligand effects on the localized surface plasmon resonance (LSPR) of gold colloids, *J. Phys. Chem. B* **108**, 13963–13971 (2004). (b) Link, S., and El-Sayed, M.A., Size and temperature dependence of the plasmon absorption of colloidal gold nanoparticles, *J. Phys. Chem. B* **103**, 4212–4217 (1999).

27. Buining, P.A., Humbel, B.M., Philipse, A.P., and Verkeij, A.J., Preparation of functional silane-stabilized gold colloids in the (sub)nanometer size range, *Langmuir* **13**, 3921–3926 (1997).

28. Templeton, A.C., Cliffel, D.E., and Murray, R.W., Redox and fluorophore function-alization of water soluble tiopronin-protected gold clusters, *J. Am. Chem. Soc.* **121**, 7081–7089 (1999).

29. Aguila, A., and Murray, R.W., Monolayer-protected clusters with fluorescent dansyl ligands, *Langmuir* **16**, 5949–5954 (2000).

30. Gu, T., Whitesell, J.K., and Fox, M.A., Energy transfer from a surface-bound arene to the gold core in ω-fluorenyl-alkane-1-thiolate monolayer-protected gold clusters, *Chem. Mater.* **15**(6), 1358–1366 (2003).

31. Hu, J., Zhang, J., Liu, F., Kittredge, K., Whitesell, J.K., and Fox, M.A., Competitive photochemical reactivity in a self-assembled monolayer on a colloidal gold cluster, *J. Am. Chem. Soc.* **123**, 1464–1470 (2001).

32. Zhang, J., Whitesell, J.K., and Fox, M.A., Photophysical behavior of variously sized colloidal gold clusters capped with monolayers of an alkylstilbenethiolate, *J. Phys. Chem. B* **107**, 6051–6055 (2003).

33. Dulkeith, E., Ringler, M., Klar, T.A., Feldmann, J., Javier, A.M., and Parak, W.J., Gold nanoparticles quench fluorescence by phase induced radiative rate suppression, *Nano Lett.* **5**, 585–589 (2005).

34. Gersten, J., and Nitzan, A., Spectroscopic properties of molecules interacting with small dielectric particles, *J. Chem. Phys.* **75**, 1139–1152 (1981).

35. Dulkeith, E., Morteani, A.C., Niedereichholz, T., Klar, T.A., Feldmann, J., Levi, S.A., van Veggel, F.C.J.M., Reinhoudt, D.N., Moller, M., and Gittins, D.I., Fluorescence quenching of dye molecules near gold nanoparticles: Radiative and nonradiative effects, *Phys. Rev. Lett.* **89**, 203002/1–203002/4 (2002).

36. Hiroshi, I., Yukiyasu, K., Takeshi, H., Yoshiyuki, E., Yoshinobu, N., Iwao, Y., and Shunichi, F., Metal and size effects on structures and photophysical properties of porphyrin-modified metal nanoclusters, *J. Mater. Chem.* **13**, 2890–2898 (2003).

37. Canepa, M., Fox, M.A., and Whitesell, J.K., The influence of core size on electronic coupling in shell-core nanoparticles: Gold clusters capped with pyrenoxylalkylthiolate, *Photochem. Photobiol. Sci.* **2**, 1177–1180 (2003).

38. Pucci, A., Tirelli, N., Willneff, E.A., Schroeder, S.L.M., Galembeck, F., and Ruggeri, G., Evidence and use of metal–chromophore interactions: Luminescence dichroism of terthiophene-coated gold nanoparticles in polyethylene oriented films, *J. Mater. Chem.* **14**, 3495–3502 (2004).

39. Ipe, B.I., Thomas, K.G., Barazzouk, S., Hotchandani, S., and Kamat, P.V., Photoinduced charge separation in a fluorophore-gold nanoassembly, *J. Phys. Chem. B* **106**, 18–21 (2002).

40. Ipe, B.I., and Thomas, K.G., Investigations on nanoparticle–chromophore and interchromophore interactions in pyrene-capped gold nanoparticles, *J. Phys. Chem. B* **108**, 13265–13272 (2004).

41. Kamat, P.V., Barazzouk, S., and Hotchandani, S., Electrochemical modulation of fluorophore emission on a nanostructured gold film, *Angew. Chem. Int. Ed. Engl.* **41**, 2764–2767 (2002).

42. Kuwahara, Y., Akiyama, T., and Yamada, S., Facile fabrication of photoelectrochemical assemblies consisting of gold nanoparticles and a tris(2,2'-bipyridine)ruthenium(II)-viologen linked thiol, *Langmuir* **17**, 5714–5716 (2001).

43. Sudeep, P.K., Ipe, B.I., Thomas, K.G., George, M.V., Barazzouk, S., Hotchandani, S., and Kamat, P.V. Fullerene-functionalized gold nanoparticles. A self-assembled photoactive antenna-metal nanocore assembly, *Nano Lett.* **2**, 29–35 (2002).

44. Shon, Y.S., and Choo, H., [60]Fullerene-linked gold nanoparticles: Synthesis and layer-by-layer growth on a solid surface, *Chem. Commun.* 2560–2561 (2002).

45. Fengjun, D., Yiyun, Y., Sungho, H., Young-Seok, S., and Shaowei, C., Fullerene-functionalized gold nanoparticles: Electrochemical and spectroscopic properties, *Anal. Chem.* **76**, 6102–6107 (2004).

46. Liu, L., Wang, T., Li, J., Guo, Z.X., Dai, L., Zhang, D., and Zhu, D., Self-assembly of gold nanoparticles to carbon nanotubes using a thiol-terminated pyrene as interlinker, *Chem. Phys. Lett.* **367**, 747–752 (2002).

47. Boal, A.K., and Rotello, V.M., Radial control of recognition and redox processes with multivalent nanoparticle hosts, *J. Am Chem. Soc* **124**, 5019–5024 (2002).

48. Yu, C.M.M., and Katz, A., Steady-state fluorescence-based investigation of the interaction between protected thiols and gold nanoparticles, *Langmuir* **18**, 2413–2420 (2002).

49. Yu, C.M.M., and Katz, A., Synthesis and characterization of gold-silica nanoparticles incorporating a mercaptosilane core–shell interface, *Langmuir* **18**, 8566–8572 (2002).

50. (a) Montalti, M., Prodi, L., Zaccheroni, N., and Battistini, G., Modulation of the photophysical properties of gold nanoparticles by accurate control of the surface coverage, *Langmuir* **20**, 7884–7886 (2004). (b) Montalti, M., Prodi, L., Zaccheroni, N., Beltrame, M., Morotti, T., and Quici, S., Stabilization of terpyridine covered gold nanoparticles by metal ions complexation, *New J. Chem.* **31**, 102–108 (2007).

51. Werts, M.H.V., Zaim, H., and Blanchard-Desce, M., Excimer probe of the binding of alkyl disulfides to gold nanoparticles and subsequent monolayer dynamics, *Photochem. Photobiol. Sci.* **3**, 29–32 (2004).

52. Montalti, M., Prodi, L., Zaccheroni, N., Baxter, R., Teobaldi, G., and Zerbetto, F., Kinetics of place-exchange reactions of thiols on gold nanoparticles, *Langmuir* **19**, 5172–5174 (2003).

53. (a) Marcaccio, M., Margotti, M., Montalti, M., Paolucci, F., Prodi, L., and Zaccheroni, N., Self-assembly of monolayer-coated silver nanoparticles on gold electrodes. An electrochemical investigation, *Collect. Czech. Chem. Commun.* **68**, 1395–1406 (2003). (b) Hicks, J.F., Zamborini, F.P., Osisek, A.J., and Murray, R.W., The dynamics of electron self-exchange between nanoparticles, *J. Am. Chem. Soc.* **123**, 7048–7053 (2001).

54. (a) Deng, Z., Tian, Y., Lee, S.-H., Ribbe, A.E., and Mao, C., DNA-encoded self-assembly of gold nanoparticles into one-dimensional arrays, *Angew. Chem. Int. Ed. Engl.* **44**, 3582–3585 (2005). (b) Guarise, C., Pasquato, L., and Scrimin, P., Reversible aggregation/deaggregation of gold nanoparticles induced by a cleavable dithiol linker, *Langmuir* **21**, 5537–5541 (2005). (c) Wessels, J.M., Nothofer, H.-G., Ford, W.E., von Wrochem, F., Scholz, F., Vossmeyer, T., Schroedter, A., Weller, H., and Yasuda, A., Optical and electrical properties of three-dimensional interlinked gold nanoparticle assemblies, *J. Am. Chem. Soc.* **126**, 3349–3356 (2004). (d) Ryan, D., Rao, S.N., Rensmo, H., Fitzmaurice, D., Preece, J.A., Wenger, S., Stoddart, J.F., and Zaccheroni, N., Heterosupramolecular chemistry: Recognition initiated and inhibited silver nanocrystal aggregation by pseudorotaxane assembly, *J. Am. Chem. Soc.* **122**, 6252–6257 (2000).

55. Liu, Y., Wang, H., Chen, Y., Ke, C.F., and Liu, M., Supramolecular aggregates constructed from gold nanoparticles and L-Try-CD polypseudorotaxanes as captors for fullerenes, *J. Am. Chem. Soc.* **127**, 657–666 (2005).

56. (a) Zhang, J., Malicka, J., Gryczynski, I., and Lakowicz, J.R., Surface-enhanced fluorescence of fluorescein-labeled oligonucleotides capped on silver nanoparticles, *J. Phys. Chem. B* **109**, 7643–7648 (2005). (b) Geddes, C.D., and Lakowicz, J. (Eds.), *Radiative Decay Engineering*, Springer, New York (2005). (c) Lukomska, J., Malicka, J., Gryczynski, I., and Lakowicz, J.R., Fluorescence enhancements on silver colloid coated surfaces, *J. Fluores.* **14**, 417–423 (2004).

57. Thomas, K.G., and Kamat, P.V., Making gold nanoparticles glow. Enhanced emission from a surface-bound fluoroprobe, *J. Am. Chem. Soc.* **122**, 2655–2656 (2000).

58. Hernàndez, F.E., Yu, S., Garcìa, M., and Campiglia, A.D., Fluorescence lifetime enhancement of organic chromophores attached to gold nanoparticles, *J. Phys. Chem. B* **19**, 9499–9504 (2005).

59. Makarova, O.V., Ostafin, A.E., Miyoshi, H., Norris, J.R., J.r., and Meisel, D., Adsorption and encapsulation of fluorescent probes in nanoparticles, *J. Phys. Chem. B* **103**, 9080–9084 (1999).

60. Chandrasekharan, N., Kamat, P.V., Hu, J., and Jones, G., II, Dye capped gold nanoclusters: Photoinduced changes in gold/rhodamine 6G nanoassemblies, *J. Phys. Chem. B* **104**, 11103–11109 (2000).

61. Barazzouk, S., Kamat, P.V., and Hotchandani, S., Photoinduced electron transfer between chlorophyll a and gold nanoparticles, *J. Phys. Chem. B* **109**, 716–723 (2005).

62. Huang, T., and Murray, R.W., Quenching of $[Ru(bpy)_3]^{2+}$ fluorescence by binding to Au nanoparticles, *Langmuir* **18**, 7077–7081 (2002).

63. Mooradian, A., Photoluminescence of metals, *Phys. Rev. Lett.* **22**, 185–187 (1969).

64. Boyd, G.T., Yu, Z.H., and Shen, Y.R., Photoinduced luminescence from the noble metals and its enhancement on roughened surfaces, *Phys. Rev. B* **33**, 7923–7936 (1986).

65. Apell, P., Monreal, R., and Lundqvist, S., Photoluminescence of noble metals, *Phys. Scr.* **38**, 174–179 (1988).

66. Wilcoxon, J.P., Martin, J.E., Parsapour, F., Wiedenman, B., and Kelley, D.F., Photoluminescence from nanosize gold clusters, *J. Chem. Phys.* **108**, 9137–9143 (1998).

67. Schaaff, T.G., Shafigullin, M.N., Khoury, J.T., Vezmar, I., Whetten, R.L., Cullen, W.G., First, P.N., Wing, C., Ascensio, J., and Yacaman, M.J., Isolation of smaller nanocrystal-Au molecules: Robust quantum effects in optical spectra, *J. Phys. Chem. B* **101**, 7885–7891 (1997).

68. Bigioni, T.P., Whetten, R.L., and Dag, O., Near-infrared luminescence from small gold nanocrystals, *J. Phys. Chem. B* **104**, 6983–6986 (2000).

69. Mohamed, M.B., Volkov, V., Link, S., and El-Sayed, M.A., The 'lightning' gold nanorods: Fluorescence enhancement of over a million compared to the gold metal, *Chem. Phys. Lett.* **317**, 517–523 (2000).

70. Huang, T., and Murray, R.W., Visible luminescence of water-soluble monolayer-protected gold clusters, *J. Phys. Chem. B* **105**, 12498–12502 (2001).

71. Link, S., Beeby, A., Fitzgerald, S., El-Sayed, M.A., Schaaff, T.G., and Whetten, R.L., Visible to infrared luminescence from a 28-atom gold cluster, *J. Phys. Chem. B* **106**, 3410–3415 (2002).

72. Link, S., El-Sayed, M.A., Schaaff, T. G., and Whetten, R.L., Transition from nanoparticle to molecular behaviour: a femtosecond transient absorption study of a size-selected 28 atom gold cluster, *Chem. Phys. Lett.* **356**, 240–246 (2002).

73. Lee, D., Donkers, R.L., Wang, G., Harper, A.S., and Murray, R.W., Electrochemistry and optical absorbance and luminescence of molecule-like Au38 nanoparticles, *J. Am. Chem. Soc.* **12**, 6193–6199 (2004).

74. Wang, G., Huang, T., Murray, R.W., Menard, L., and Nuzzo, R.G., Near-IR luminescence of monolayer-protected metal clusters, *J. Am. Chem. Soc.* **127**, 812–813 (2005).

6

Optics of Slanted Chiral STFs

Fei Wang

Micron Technology, Inc., 8000 S. Federal Way, P. O. Box 6, Boise, ID 83707–0006

6.1 Introduction

6.1.1 Sculptured Thin Films

6.1.1.1 A General Picture

The concept of sculptured thin films (STFs) and an associated technology for optics emerged during the 1990s from the widely used columnar thin films (CTFs) [1–3]. The ideal morphology of CTFs consists of almost identical, straight, and parallel nanowires of elliptical cross section. The nanowires in an STF are not straight, but are made to grow as curves that veer away from the substrate [4–6], as shown in Figure 6.1. The nanowire diameters range from about 10 to 100 nm, and a wide variety of such morphologies are realizable through instantaneous variation in the growth direction of nanowires during physical vapor deposition (PVD) [7, 8].

Two canonical classes of STF morphologies — nematic and helicoidal — are obtainable by choosing the proper axis of rotation of the substrate during PVD. Nematic morphologies are two-dimensional (2D), ranging from the simple chevrons to the more complex C and S shapes. Helicoidal morphologies are three–dimensional (3D), including helical and superhelical [9]. More complex morphologies and multisection STFs, in which the chemical composition or/and the nanowire shape varies from section to section along the thickness direction, can also be realized on large–area substrates [10, 11].

A wide variety of materials can be used for PVD of STFs, ranging from insulators (e.g., oxides and fluorides) to semiconductors (e.g., crystalline silicon) to metals (e.g., aluminum and chromium). This diversity reflects the low degree of sensitivity of morphology obtained by PVD to chemical composition. In fact, as predicted by the structure zone model of Thornton [12], the morphology obtained through PVD is largely a consequence of obeying simple geometric rules of atomic aggregation, when little or no surface diffusion is involved — which explains why essentially similar morphologies can be generated from a broad range of material sources.

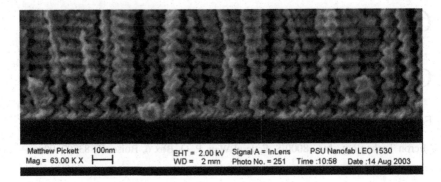

Matthew Pickett   100nm   EHT = 2.00 kV   Signal A = InLens   PSU Nanofab LEO 1530
Mag = 63.00 K X ⊢——⊣   WD =   2 mm   Photo No. = 251   Time :10:58   Date :14 Aug 2003

FIGURE 6.1. Scanning electron micrograph of a chiral STF of silicon oxide. This thin film is an assembly of parallel helical nanowires. (Courtesy: Mark W. Horn, Pennsylvania State University.)

Being porous, STFs contain voids of characteristic shapes and sizes. As these voids can be filled with different materials, STFs can function in many different ways. For example, the fabrication of low–permittivity nanocomposites for the microelectronics industry has been suggested using STF technology [13], and proof–of–concept optical fluid sensors have been designed and fabricated to exploit the textured porosity of STFs [14, 15].

In order to fully harness STF technology, an electromagnetic model of STFs needs to be established with the capability to account for structure–property relationships. In the macroscopic sense, an STF is a unidirectionally nonhomogeneous, bianisotropic continuum in the optical regime. In the microscopic sense, an STF is viewed as a composite material with at least two different material phases molded into the sculptured morphology. The relationship between the nanostructure and the macroscopic constitutive properties needs to be quantitatively delineated and understood. For that purpose, a nominal model has been developed to determine the constitutive dyadics of STFs from local homogenization of electrically small ellipsoidal particles of different material phases [16–19]. Intelligent design and fabrication of STF devices are made possible by coupling this nominal model with experimentation [18].

6.1.1.2 Growth of STFs

In general, STFs are fabricated by directional PVD methods, such as evaporation in high–vacuum conditions or by sputtering at intermediate–vacuum conditions. The schematic of a basic system for PVD of STFs on planar substrates is presented in Figure 6.2. There are two fundamental axes of rotation of the substrate: one (z axis) is perpendicular to the substrate plane (xy plane), and the other (y axis) is parallel to the substrate plane. The incident vapor flux makes an oblique angle χ_v to the substrate plane.

The growth of a STF starts with the process of nucleation and involves continuous renucleation thereafter. These processes are controlled through the

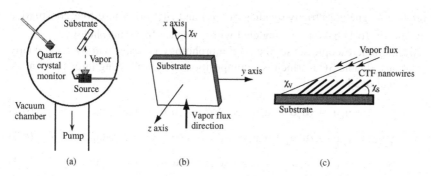

FIGURE 6.2. (a) Schematic of the basic system for PVD of STFs on planar substrates. (b) Coordinate system with the y and the z axes indicated as the two fundamental axes of rotation of the substrate. (c) Vapor incidence angle χ_v and the tilt angle χ_s.

proper choice of PVD parameters — such as energy of bombarding particles, substrate temperature, and χ_v. Once the growth reaches a steady state, the nanowire shapes can be tailored by controlling the rotation of substrate. For example, nematic STFs are attained by rotating the substrate about the y axis [20], while helicoidal STFs are formed with the substrate rotation about the z axis [21, 22]. The mass density of helicoidal STFs is expected not to vary in the thickness direction since χ_v is fixed during growth, so long as the nanowires attain a steady–state diameter in the early nucleation and growth stages. By changing the rotational speed and the orientation of the substrate sequentially without halting deposition, multisection STFs have been fabricated with cascaded morphologies in the thickness direction [10, 11].

6.1.1.3 Canonical Delineation of STFs

The macroscopic conception of STFs in the optical regime is as unidirectionally nonhomogeneous, bianisotropic continuums, with the linear constitutive relations

$$\left.\begin{aligned}
\mathbf{D}(\mathbf{r}, \omega) &= \epsilon_0[\underline{\underline{\epsilon}}(z, \omega) \cdot \mathbf{E}(\mathbf{r}, \omega) + \underline{\underline{\alpha}}(z, \omega) \cdot \mathbf{H}(\mathbf{r}, \omega)] \\
\mathbf{B}(\mathbf{r}, \omega) &= \mu_0[\underline{\underline{\beta}}(z, \omega) \cdot \mathbf{E}(\mathbf{r}, \omega) + \underline{\underline{\mu}}(z, \omega) \cdot \mathbf{H}(\mathbf{r}, \omega)]
\end{aligned}\right\} \tag{6.1}$$

indicating that the axis of nonhomogeneity is parallel to the z axis. In (6.1) and hereafter, $\mathbf{E}(\mathbf{r}, \omega)$ and $\mathbf{H}(\mathbf{r}, \omega)$ are the electric and magnetic field phasors, respectively; $\mathbf{D}(\mathbf{r}, \omega)$ and $\mathbf{B}(\mathbf{r}, \omega)$ are the electric and magnetic displacement phasors, respectively; an $\exp(-i\omega t)$ time–dependence is implicit, with $i = \sqrt{-1}$, ω as the angular frequency of light, and t as time; $\mathbf{r} = x\mathbf{u}_x + y\mathbf{u}_y + z\mathbf{u}_z$ is the position vector, with $\mathbf{u}_x$, $\mathbf{u}_y$, and $\mathbf{u}_z$ as unit Cartesian vectors; and $\epsilon_0 = 8.854 \times 10^{-12}$ F m^{-1} and $\mu_0 = 4\pi \times 10^{-7}$ H m^{-1} are the permittivity and permeability of free space, respectively. The electric and magnetic properties of STFs are expressed through the relative permittivity dyadic $\underline{\underline{\epsilon}}(z, \omega)$ and the relative permeability dyadic $\underline{\underline{\mu}}(z, \omega)$, respectively; the magnetoelectric properties are expressed through the relative magnetoelectricity dyadics $\underline{\underline{\alpha}}(z, \omega)$ and $\underline{\underline{\beta}}(z, \omega)$. These four constitutive dyadics in (6.1) have to be modeled with guidance from morphology.

In the macroscopic sense, the STF morphology is delineated by a uniaxially rotational dyadic $\underline{\underline{S}}(z)$, which is a composition of the following three elementary rotational dyadics:

$$\underline{\underline{S}}_x(z) = \mathbf{u}_x\mathbf{u}_x + (\mathbf{u}_y\mathbf{u}_y + \mathbf{u}_z\mathbf{u}_z)\cos\zeta_x(z) + (\mathbf{u}_z\mathbf{u}_y - \mathbf{u}_y\mathbf{u}_z)\sin\zeta_x(z), \tag{6.2}$$

$$\underline{\underline{S}}_y(z) = \mathbf{u}_y\mathbf{u}_y + (\mathbf{u}_x\mathbf{u}_x + \mathbf{u}_z\mathbf{u}_z)\cos\zeta_y(z) + (\mathbf{u}_z\mathbf{u}_x - \mathbf{u}_x\mathbf{u}_z)\sin\zeta_y(z), \tag{6.3}$$

$$\underline{\underline{S}}_z(z) = \mathbf{u}_z\mathbf{u}_z + (\mathbf{u}_x\mathbf{u}_x + \mathbf{u}_y\mathbf{u}_y)\cos\zeta_z(z) + (\mathbf{u}_y\mathbf{u}_x - \mathbf{u}_x\mathbf{u}_y)\sin\zeta_z(z). \tag{6.4}$$

Here, $\zeta_x(z)$, $\zeta_y(z)$, and $\zeta_z(z)$ are angular functions indicating rotations about the x, y, and z axes, respectively. Accordingly, the linear constitutive relations of a single–section STF are set up as follows:

$$\left. \begin{aligned} \mathbf{D}(\mathbf{r}, \omega) &= \epsilon_0\underline{\underline{S}}(z) \cdot [\underline{\underline{\epsilon}}_{ref}(\omega) \cdot \underline{\underline{S}}^T(z) \cdot \mathbf{E}(\mathbf{r}, \omega) + \underline{\underline{\alpha}}_{ref}(\omega) \cdot \underline{\underline{S}}^T(z) \cdot \mathbf{H}(\mathbf{r}, \omega)] \\ \mathbf{B}(\mathbf{r}, \omega) &= \mu_0\underline{\underline{S}}(z) \cdot [\underline{\underline{\beta}}_{ref}(\omega) \cdot \underline{\underline{S}}^T(z) \cdot \mathbf{E}(\mathbf{r}, \omega) + \underline{\underline{\mu}}_{ref}(\omega) \cdot \underline{\underline{S}}^T(z) \cdot \mathbf{H}(\mathbf{r}, \omega)] \end{aligned} \right\} . \tag{6.5}$$

The dyadics $\underline{\underline{\epsilon}}_{ref}(\omega)$, etc., are termed *reference* relative constitutive dyadics that are spatially invariant, and the superscript T denotes the transpose. The constitutive equations (6.5) reflect the fact that the morphologies of a single–section STF in any two planes $z = z_1$ and $z = z_2$ are exactly interchangeable by a suitable rotation.

Canonical forms of STFs have been delineated. For example, by choosing $\underline{\underline{S}}(z) = \underline{\underline{S}}_y(z)$, (6.5) describes the canonical class of STFs with nematic morphology. The choice of $\underline{\underline{S}}(z) = \underline{\underline{S}}_z(z)$ gives rise to the canonical class of STFs with helicoidal morphology. Helicoidal STFs are exemplified by chiral STFs for which $\zeta_z(z) = \pm\pi z/\Omega$ in (6.4), where 2Ω is the structural period, and the $+/-$ signs indicate the structural right/left–handedness of the film. More complicated specifications of $\underline{\underline{S}}(z)$ are possible, as for the STFs with superhelical morphology [9].

Because of the inclination of the nanowires with respect to the z axis, it is appropriate to delineate $\underline{\underline{\epsilon}}_{ref}(\omega)$, etc., in terms of the reference unit vectors

$$\left. \begin{aligned} \mathbf{u}_\tau &= \mathbf{u}_x\cos\chi_s + \mathbf{u}_z\sin\chi_s \\ \mathbf{u}_n &= -\mathbf{u}_x\sin\chi_s + \mathbf{u}_z\cos\chi_s \\ \mathbf{u}_b &= -\mathbf{u}_y \end{aligned} \right\}, \tag{6.6}$$

where χ_s is called the tilt angle. The specification

$$\underline{\underline{\sigma}}_{ref}(\omega) = \sigma_a(\omega)\mathbf{u}_n\mathbf{u}_n + \sigma_b(\omega)\mathbf{u}_\tau\mathbf{u}_\tau + \sigma_c(\omega)\mathbf{u}_b\mathbf{u}_b, \quad \sigma \in (\epsilon, \alpha, \beta, \mu), \tag{6.7}$$

is in accord with the local orthorhombicity of STFs. In fact, if the nanowire cross section is circular, $\sigma_a(\omega) = \sigma_c(\omega)$ and the unit vectors $\mathbf{u}_n$ and $\mathbf{u}_b$ can be chosen arbitrarily in the cross–section plane of the nanowire. When the nanowire cross section is elliptical, $\sigma_a(\omega) \neq \sigma_c(\omega)$ and the unit vectors $\mathbf{u}_n$ and $\mathbf{u}_b$ should be along the two principal axes of the cross–section ellipse. More generally, the nanowire cross section can be of any convex shape defined by

$$\mathbf{r}_s = \underline{\underline{U}}(\theta) \cdot \mathbf{u}_r(\theta). \tag{6.8}$$

Here, $\mathbf{u}_r(\theta)$ is the radial unit vector in a two-dimensional polar coordinate system and $\underline{\underline{U}}(\theta)$ is the 2×2 shape dyadic which is assumed to be real symmetric because of the convexity of the cross section. Clearly, the determination of the in–plane dyadic $\underline{\underline{\sigma}}^{\parallel}_{ref}(\omega)$ — which is the projection of $\underline{\underline{\sigma}}_{ref}(\omega)$ in the nanowire cross–section plane — through (6.8) can be achieved by a local homogenization procedure (e.g., Bruggeman formalism) detailed elsewhere [16–19]. However, because $\underline{\underline{U}}(\theta)$ is real symmetric, it can be shown that $\underline{\underline{\sigma}}^{\parallel}_{ref}(\omega)$ should be a symmetric 2×2 dyadic which can be diagonalized as

$$\underline{\underline{\sigma}}^{\parallel}_{ref}(\omega) = \underline{\underline{V}}(\omega) \cdot \underline{\underline{\sigma}}^{\parallel}_{diag}(\omega) \cdot \underline{\underline{V}}^T(\omega), \quad \sigma \in (\epsilon, \alpha, \beta, \mu). \tag{6.9}$$

Here, $\underline{\underline{\sigma}}^{\parallel}_{diag}(\omega)$ is the 2×2 diagonal dyadic whose two nonzero entries determine the values of $\sigma_a(\omega)$ and $\sigma_c(\omega)$; and $\underline{\underline{V}}(\omega)$ is the 2×2 orthogonal dyadic comprising the normalized eigenvectors of $\underline{\underline{\sigma}}^{\parallel}_{ref}(\omega)$. By choosing the in–plane unit vectors $\mathbf{u}_n$ and $\mathbf{u}_b$ as the two normalized eigenvectors contained in $\underline{\underline{V}}(\omega)$, $\underline{\underline{\sigma}}_{ref}(\omega)$ can still be expressed as per (6.7), which is tantamount to the local orthorhombicity of STFs in general. By choosing $\sigma_a(\omega) \neq \sigma_b(\omega) \neq \sigma_c(\omega)$ in general for $\underline{\underline{\sigma}}_{ref}(\omega)$, the density anisotropy occurring during PVD is thus also taken into account [1].

For magneto–optics, gyrotropic terms such as $i\sigma_g(\omega)\mathbf{u}_\tau \times \underline{\underline{I}}$ can be included in (6.7), where $\underline{\underline{I}}$ is the identity dyadic [23].

The Post constraint

$$\text{Trace}\left\{\underline{\underline{\mu}}^{-1}_{ref}(\omega)\left[\underline{\underline{\beta}}_{ref}(\omega) + \frac{\epsilon_0}{\mu_0}\underline{\underline{\alpha}}_{ref}(\omega)\right]\right\} \equiv 0 \tag{6.10}$$

is mandated by the Lorentz–Heaviside visualization of electromagnetic theory [24]. If $\underline{\underline{\epsilon}}_{ref}(\omega) = \underline{\underline{\epsilon}}^T_{ref}(\omega)$, $\underline{\underline{\mu}}_{ref}(\omega) = \underline{\underline{\mu}}^T_{ref}(\omega)$, and $\underline{\underline{\alpha}}_{ref}(\omega) = -\underline{\underline{\beta}}^T_{ref}(\omega)$, the STF is Lorentz–reciprocal [25]. The simplest form of STFs is the purely dielectric one, so that $\underline{\underline{\mu}}_{ref}(\omega) = \underline{\underline{I}}$, whereas both $\underline{\underline{\alpha}}_{ref}(\omega)$ and $\underline{\underline{\beta}}_{ref}(\omega)$ are null dyadics.

6.1.1.4 From Nanostructure to Continuum

The constitutive relations (6.5) are set by viewing any STF as a continuous medium. The nanoscale information regarding morphology and composition needs to be reflected in this macroscopic model. As any STF is a composite material with at least two different material phases, the constitutive dyadics $\underline{\underline{\epsilon}}(z, \omega)$, etc., can be modeled through the commonly used procedure of homogenization, so long as the particulate dimensions of all material phases are much smaller than the wavelength of incident light [26]. However, as a STF is effectively nonhomogeneous in the z direction, the homogenization procedure must be implemented in the localized fashion, i.e., for any thin slice of the STF perpendicular to the z axis. But any two slices of a single–section STF are in fact identical, except for a rotational transformation captured by $\underline{S}(z)$. Therefore, the local homogenization procedure for the STF can be performed for $\underline{\underline{\epsilon}}_{ref}(\omega)$, etc., but leading to the construction of $\underline{\underline{\epsilon}}(z, \omega)$, etc.

In a nominal model that has been developed during the last five years [16–19], the nanowires as well as interparticle voids/fillings in a STF are represented as parallel strings of electrically small ellipsoidal particles. Both ellipsoidal shape factors and the volumetric proportions of the material phases must be chosen for the implementation of local homogenization procedure. Once an algorithm for the macroscopic properties of the STF has been set up, calibration of this nominal model is possible by comparison of the predicted optical responses against measured data [18].

6.1.2 Chiral STFs

Chiral STFs are a subclass of helicoidal STFs that are periodically nonhomogeneous along the z axis. It is easy to fabricate them with periods specified between 50 nm and 2 μm. A chiral STF is described by specifying $\underline{S}(z) = \underline{S}_z(z)$ in (6.5), along with $\zeta_z(z) = \pm \pi z / \Omega$. Modeled as purely dielectric substances, these have been the most popular STFs for optics to date.

6.1.2.1 Circular Bragg Phenomenon

The attraction of chiral STFs is attributed to the circular Bragg phenomenon evinced by them [9]. Briefly, a structurally right/left–handed chiral STF only a few periods thick reflects almost completely normally incident right/left circularly polarized (RCP/LCP) plane waves with wavelength lying in the so–called Bragg regime; while the reflection of normally incident LCP/RCP plane waves in the same regime is very little.

Certainly, the circular Bragg phenomenon occurs for oblique incidence as well, but it is greatly influenced by the directionality of planewave incidence [27, 28]. For example, the Bragg regime shifts to shorter wavelengths as the polar angle of planewave incidence θ_i^p (with respect to the z axis) increases in absolute value. The width of the Bragg regime also decreases with increasing $|\theta_i^p|$. The azimuthal angle of planewave incidence ψ_i^p (with respect to the x axis in the xy plane) does not affect the Bragg regime strongly, but it does significantly affect properties of optical rotation and ellipticity in the Bragg regime. There is also a decrease in discrimination between LCP and RCP plane waves when $|\theta_i^p|$ is very large [29].

A simple explanation of circular Bragg phenomenon for chiral STFs is provided by the application of coupled–wave theory (CWT) for normal incidence [30–32]: When the incident wavelength is approximately equal to the optical period of the dielectric nonhomogeneity, the morphology of a chiral STF acts as a scalar Bragg grating for co–handed circularly polarized (CP) plane waves, but not for cross–handed circularly polarized plane waves. Rather, the chiral STF acts as a homogeneous, isotropic, dielectric medium in the latter case. This discrimination of circularly polarized plane waves by chiral STFs can be harnessed in many different ways, so that optical applications of chiral STFs are abundant.

6.1.2.2 Optical Applications

Many optical applications of chiral STFs have been proposed and even realized to date. By utilizing the circular Bragg phenomenon, chiral STFs as circular polarization filters have been theoretically examined and then experimentally realized [9, 33]. A handedness–selective light inverter, which comprises a chiral STF and a CTF functioning as a half waveplate, was also designed and then fabricated and tested [34, 35]. By introducing either a layer defect or a twist defect in the middle of a chiral STF, narrowband spectral–hole filters have been designed and implemented [15, 36, 37]. The operational free–space wavelength of these filters is located in the Bragg regime for normal incidence, and is dependent on the nature of the defect.

Piezoelectrically tunable lasers made of dye–doped polymer chiral STFs have been recently proposed as a result of theoretical analysis of piezoelectric manipulation of chiral STFs [38, 39]. Time–domain exhibition of the circular Bragg phenomenon has been studied, which would lead to the use of chiral STFs to shape optical pulses in optical communication systems [40, 41]. Being porous, chiral STFs are also useful for optically sensing humidity and various chemicals [15, 42, 43].

6.1.3 Slanted Chiral STFs

6.1.3.1 Genesis

The helical nanowires of a chiral STF stand upright on a substrate. Therefore, the optical periodicity of a chiral STF is unidirectional — along the normal to the substrate plane, i.e., the z axis. A chiral STF can be viewed as a volume grating that consists of infinitesimally thin dielectric sublayers. The constitutive properties are homogeneous within each sublayer, but vary periodically from sublayer to sublayer to form a chiral architecture. The periodicity of the chiral architecture implies that a chiral STF is indeed a volume grating, albeit different from the ones usually studied [44].

There exists another type of gratings of great importance in optics, however. These are diffraction gratings [45]. The distinguishing feature of a diffraction grating is the periodic variation of constitutive properties in a plane, often achieved by periodically corrugating a planar sheet. As a consequence of illumination by a plane wave, the reflected, the refracted, and the transmitted fields of a diffraction grating are discrete angular spectrums of propagating as

well as evanescent plane waves — called the Floquet harmonics, due to their
emergence from the Floquet–Bloch theorem [46]. As either the frequency or
the direction of incidence is altered, an evanescent Floquet harmonic of the
reflected/refracted/transmitted field may turn into a propagating one or vice versa,
this phenomenon being manifested as the so–called Rayleigh–Wood anomaly
[45] in the remittance (i.e., reflectance and transmittance) spectrums.

Clearly, the coupling of these two types of gratings (i.e., volume and
diffraction gratings) is likely to be rich in optical phenomena and applications,
the more so when the volume grating is sensitive to circular polarization.
A direct outcome of the coupling is that the circular Bragg phenomenon would
be affected by the nonspecular angular spectrums of diffraction as well as by
the associated Rayleigh–Wood anomalies. In order to physically achieve this
coupling, slanted chiral STFs were conceptualized for this thesis as an extension
of chiral STFs [47].

Morphologically, a slanted chiral STF comprises helical nanowires slanted at
an angle $\alpha \neq 0$ to the normal to the substrate plane, as schematically illustrated in
Figure 6.3. A possible way to attain this morphology is by rotating the substrate
with a variable angular velocity during deposition [48, 49]. A slanted chiral STF
will thus be unidirectionally and periodically nonhomogeneous along an axis
inclined, but not perpendicular to the substrate plane.

Therefore, a slanted chiral STF is periodic along the z and the x axes as well.
In order to represent the slanted morphology, $\underline{\underline{S}}(z)$ in (6.5) is replaced by

$$\underline{\underline{S}}(\mathbf{r}) = \underline{\underline{s}}_y(-\alpha) \cdot \underline{\underline{s}}_z(\mathbf{r}),\qquad(6.11)$$

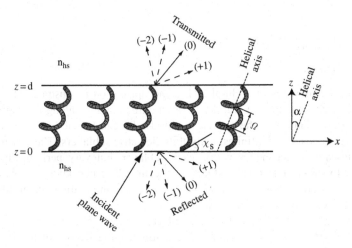

FIGURE 6.3. Schematic of the boundary value problem involving a slanted chiral STF.
Specular reflection and transmission carry the label $n = 0$ and are identified by solid
arrows, while nonspecular reflection and transmission carry the labels $n \neq 0$ and are
identified by dashed arrows.

where

$$\underline{\underline{s}}_y(\alpha) = \mathbf{u}_y\mathbf{u}_y + (\mathbf{u}_x\mathbf{u}_x + \mathbf{u}_z\mathbf{u}_z)\cos\alpha + (\mathbf{u}_z\mathbf{u}_x - \mathbf{u}_x\mathbf{u}_z)\sin\alpha \qquad (6.12)$$

is similar to $\underline{\underline{S}}_y(z)$, while

$$\underline{\underline{s}}_z(\mathbf{r}) = \mathbf{u}_z\mathbf{u}_z + (\mathbf{u}_x\mathbf{u}_x + \mathbf{u}_y\mathbf{u}_y)\cos\left[\frac{\pi}{\Omega}(\mathbf{r}\cdot\mathbf{u}_\ell)\right] + h(\mathbf{u}_y\mathbf{u}_x - \mathbf{u}_x\mathbf{u}_y)\sin\left[\frac{\pi}{\Omega}(\mathbf{r}\cdot\mathbf{u}_\ell)\right]$$

$$(6.13)$$

is similar to $\underline{\underline{S}}_z(z)$ except that the axis of nonhomogeneity is parallel to the unit vector

$$\mathbf{u}_\ell = \sin\alpha\,\mathbf{u}_x + \cos\alpha\,\mathbf{u}_z. \qquad (6.14)$$

Therefore, $\underline{\underline{s}}_z(\mathbf{r})$ in (6.13) is a function of both x and z, and the structural handedness is specified through the parameter $h = \pm 1$.

The constitutive relations of a single–section slanted chiral STF are thereby written as

$$\left.\begin{array}{l}
\mathbf{D}(\mathbf{r}, \omega) = \epsilon_0\underline{\underline{s}}_y(-\alpha) \cdot \underline{\underline{s}}_z(\mathbf{r}) \cdot \left[\underline{\underline{\epsilon}}_{ref}(\omega) \cdot \underline{\underline{s}}_z^{\mathrm{T}}(\mathbf{r}) \cdot \underline{\underline{s}}_y^{\mathrm{T}}(-\alpha) \cdot \mathbf{E}(\mathbf{r}, \omega)\right. \\
\qquad\qquad \left. + \underline{\underline{\alpha}}_{ref}(\omega) \cdot \underline{\underline{s}}_z^{\mathrm{T}}(\mathbf{r}) \cdot \underline{\underline{s}}_y^{\mathrm{T}}(-\alpha) \cdot \mathbf{H}(\mathbf{r}, \omega)\right] \\[2mm]
\mathbf{B}(\mathbf{r}, \omega) = \mu_0\underline{\underline{s}}_y^{\mathrm{T}}(-\alpha) \cdot \underline{\underline{s}}_z(\mathbf{r}) \cdot \left[\underline{\underline{\beta}}_{ref}(\omega) \cdot \underline{\underline{s}}_z^{\mathrm{T}}(\mathbf{r}) \cdot \underline{\underline{s}}_y^{\mathrm{T}}(-\alpha) \cdot \mathbf{E}(\mathbf{r}, \omega)\right. \\
\qquad\qquad \left. + \underline{\underline{\mu}}_{ref}(\omega) \cdot \underline{\underline{s}}_z^{\mathrm{T}}(\mathbf{r}) \cdot \underline{\underline{s}}_y^{\mathrm{T}}(-\alpha) \cdot \mathbf{H}(\mathbf{r}, \omega)\right]
\end{array}\right\} . \qquad (6.15)$$

Clearly from (6.12) to (6.15), a slanted chiral STF couples the two types of gratings in its constitutive properties, and the slant angle $\alpha \neq 0$ totally controls this coupling. In the remainder of this chapter, $\underline{\underline{\mu}}_{ref}(\omega) = \underline{\underline{I}}$ while $\underline{\underline{\alpha}}_{ref}(\omega)$ and $\underline{\underline{\beta}}_{ref}(\omega)$ are null dyadics, for simplicity as well as in accord with the dielectric nature of most optical films.

6.1.3.2 Optics of Slanted Chiral STFs

The basic feature of slanted chiral STFs is likely to be visualized best in terms of the planewave response [50, 51]. The circular Bragg phenomenon may occur nonspecularly due to the dual periodicity in morphology. Affected by the slant angle α, the circular Bragg phenomenon could become even tunable in both spectrum and direction; and many different optical applications could arise [51, 52].

Furthermore, structural defects parallel to the substrate plane could be easily incorporated into the nanostructure of slanted chiral STFs [53, 54]. These structural defects generally produce a wave resonance localized inside the Bragg regime or at its edges. Both frequency and polarization characteristics of wave resonance have been harnessed for the design of narrowband optical filters in chiral STFs [36, 37], and low–threshold lasers and low–loss waveguides are also possible for being devised therefrom [55–57]. There is a crossover phenomenon associated with the localization of wave resonance in both chiral STFs and cholesteric liquid crystals (CLCs) [58–63]: The localization is seen as a co–handed reflectance hole when the chiral structure is relatively thin, but as a cross–handed transmission hole when the thickness is large. No doubt, this phenomenon would also be found in slanted chiral STFs in the presence of some types of defects. Exploitation of the crossover phenomenon for new devices is also possible, as exemplified by the design of a superior filter by combining both layer and twist defects in a chiral STF [58].

This chapter is organized as follows: Section 6.2 describes the planewave response of a slanted chiral STF under normal–incidence and oblique–incidence conditions. Section 6.3 presents the phenomenon of wave resonance in a slanted chiral STF with a central twist defect, and Section 6.4 explains the defect–mode crossover phenomenon in an analytical approach. Concluding remarks are summarized in Section 6.5.

6.2 Response of Slanted Chiral STFs to Plane Waves

6.2.1 Geometry of the Basic Problem

Let the region $0 \leq z \leq d$ be occupied by a slanted chiral STF, as shown in Figure 6.3, while the half–spaces $z \leq 0$ and $z \geq d$ are filled with a homogeneous, isotropic, and dielectric medium of refractive index n_{hs}. The general constitutive relations of a single–section slanted chiral STF have been specified in (6.15). However, the focus hereafter is on dielectric slanted chiral STFs for which the relative permittivity dyadic

$$\underline{\underline{\epsilon}}(\mathbf{r}, \omega) = \underline{\underline{s}}_y(-\alpha) \cdot \underline{\underline{s}}_z(\mathbf{r}) \cdot \underline{\underline{\epsilon}}_{ref}(\omega) \cdot \underline{\underline{s}}_z^{\mathrm{T}}(\mathbf{r}) \cdot \underline{\underline{s}}_y^{\mathrm{T}}(-\alpha). \qquad (6.16)$$

The reference relative permittivity dyadic $\underline{\underline{\epsilon}}_{ref}(\omega)$ in (6.16) can be visualized either from a microscopic perspective or from a phenomenological perspective. Microscopic modeling of $\underline{\underline{\epsilon}}_{ref}(\omega)$ takes account of the nanostructure, and links it to macroscopic observables. As exemplified by a nominal model for chiral STFs [16–19], $\underline{\underline{\epsilon}}_{ref}(\omega)$ can be derived from a local homogenization procedure by viewing an STF as an ensemble of oriented ellipsoidal particles of different material phases. The Bruggeman formalism is used to implement a local homogenization procedure in this model, but other formalisms can also be used [26].

From the macroscopic point of view, a dielectric STF is locally orthorhombic in most cases [6, 9]. Therefore, $\underline{\underline{\epsilon}}_{ref}(\omega)$ is set up as

$$\underline{\underline{\epsilon}}_{ref}(\omega) = \epsilon_a \mathbf{u}_n \mathbf{u}_n + \epsilon_b \mathbf{u}_\tau \mathbf{u}_\tau + \epsilon_c \mathbf{u}_b \mathbf{u}_b. \qquad (6.17)$$

Most simply, the scalars $\epsilon_{a,b,c}(\omega)$ are assumed to emerge from a single–resonance Lorentzian model [64, 65] such that

$$\epsilon_\sigma(\omega) = 1 + \frac{p_\sigma}{\left[1 + (N_\sigma^{-1} - i\omega_\sigma^{-1}\omega)^2\right]}, \qquad \sigma = a, b, c. \tag{6.18}$$

Here, the parameters $p_{a,b,c}$ are the oscillator strengths, while $\omega_{a,b,c}$ and $N_{a,b,c}$ determine the resonance frequencies and absorption linewidths. For later convenience, the composite scalar

$$\epsilon_d = \frac{\epsilon_a \epsilon_b}{\cos^2\chi_s \epsilon_a + \sin^2\chi_s \epsilon_b} \tag{6.19}$$

is introduced, which, combined with ϵ_c, suffices for optical investigations of axially excited chiral STFs [30–33].

For slanted chiral STFs, the restriction $|\alpha| < \chi_s$ should be made because the nanowires must always grow upwards in relation to the substrate plane [47]. Specially, $\alpha = 0$ specifies chiral STFs with the consequent reduction $\underline{\epsilon}(\mathbf{r}, \omega) \to \underline{\epsilon}(z, \omega)$ in (6.16). From here onwards, the dependences of various quantities on ω are implicit. Instead, the dependences on the free–space wavelength $\lambda_0 = 2\pi/\omega\sqrt{\mu_0\epsilon_0}$ may be written explicitly for emphasis.

6.2.2 Electromagnetic Wave Propagation

6.2.2.1 Field Representation

A plane wave is incident from the half–space $z \leq 0$ onto the plane $z = 0$. As a result, reflection and transmission into the two half–spaces occur. Let the incident plane wave propagate with the wavevector $\mathbf{k}_+^{(0)} = k_x^{(0)}\mathbf{u}_x + k_y^{(0)}\mathbf{u}_y + k_z^{(0)}\mathbf{u}_z$. The incident, reflected, and transmitted electromagnetic field phasors are expressed in sets of Floquet harmonics respectively as [50]

$$\mathbf{E}_i = \sum_{n\in\mathbb{Z}} \left(\mathbf{L}_+^{(n)} a_L^{(n)} + \mathbf{R}_+^{(n)} a_R^{(n)}\right) \exp\left(i\mathbf{k}_+^{(n)}\cdot\mathbf{r}\right), \qquad z \leq 0, \tag{6.20}$$

$$\mathbf{H}_i = \frac{-in_{hs}}{\eta_0} \sum_{n\in\mathbb{Z}} \left(\mathbf{L}_+^{(n)} a_L^{(n)} - \mathbf{R}_+^{(n)} a_R^{(n)}\right) \exp\left(i\mathbf{k}_+^{(n)}\cdot\mathbf{r}\right), \qquad z \leq 0, \tag{6.21}$$

$$\mathbf{E}_r = \sum_{n\in\mathbb{Z}} \left(\mathbf{L}_-^{(n)} r_L^{(n)} + \mathbf{R}_-^{(n)} r_R^{(n)}\right) \exp\left(i\mathbf{k}_-^{(n)}\cdot\mathbf{r}\right), \qquad z \leq 0, \tag{6.22}$$

$$\mathbf{H}_r = \frac{-in_{hs}}{\eta_0} \sum_{n\in\mathbb{Z}} \left(\mathbf{L}_-^{(n)} r_L^{(n)} - \mathbf{R}_-^{(n)} r_R^{(n)}\right) \exp\left(i\mathbf{k}_-^{(n)}\cdot\mathbf{r}\right), \qquad z \leq 0, \tag{6.23}$$

$$\mathbf{E}_t = \sum_{n\in\mathbb{Z}} \left(\mathbf{L}_+^{(n)} t_L^{(n)} + \mathbf{R}_+^{(n)} t_R^{(n)}\right) \exp\left(i\mathbf{k}_+^{(n)}\cdot\tilde{\mathbf{r}}\right), \qquad z \geq d, \tag{6.24}$$

$$\mathbf{H}_t = \frac{-in_{hs}}{\eta_0} \sum_{n\in\mathbb{Z}} \left(\mathbf{L}_+^{(n)} t_L^{(n)} - \mathbf{R}_+^{(n)} t_R^{(n)}\right) \exp\left(i\mathbf{k}_+^{(n)}\cdot\tilde{\mathbf{r}}\right), \qquad z \geq d. \tag{6.25}$$

In (6.20)–(6.25) and hereafter, $\eta_0 = \sqrt{\mu_0/\epsilon_0}$ is the intrinsic impedance of free space; while $\left\{a_L^{(n)}, a_R^{(n)}\right\}$, $\left\{r_L^{(n)}, r_R^{(n)}\right\}$, and $\left\{t_L^{(n)}, t_R^{(n)}\right\}$, respectively, are complex–valued amplitudes of the LCP and RCP components of the Floquet harmonic of order n of the incident, reflected, and transmitted fields. The symbol $\mathbb{Z}$ represents the set $\{0, \pm1, \pm2, \ldots\}$ of all integers, and $\tilde{\mathbf{r}} = \mathbf{r} - d\mathbf{u}_z$.

The wavevectors $\mathbf{k}_\pm^{(n)}$ as well as the circular polarization vectors $\mathbf{L}_\pm^{(n)}$ and $\mathbf{R}_\pm^{(n)}$ of the Floquet harmonic of order n are specified as

$$\mathbf{k}_\pm^{(n)} = k_x^{(n)}\mathbf{u}_x + k_y^{(0)}\mathbf{u}_y \pm k_z^{(n)}\mathbf{u}_z \,, \tag{6.26}$$

$$\mathbf{L}_\pm^{(n)} = \pm\left(i\mathbf{s}^{(n)} - \mathbf{p}_\pm^{(n)}\right)/\sqrt{2}\,, \tag{6.27}$$

$$\mathbf{R}_\pm^{(n)} = \mp\left(i\mathbf{s}^{(n)} + \mathbf{p}_\pm^{(n)}\right)/\sqrt{2}\,. \tag{6.28}$$

In these expressions, the vectors

$$\left.\begin{aligned}
\mathbf{s}^{(n)} &= \frac{-k_y^{(0)}}{k_{xy}^{(n)}}\mathbf{u}_x + \frac{k_x^{(n)}}{k_{xy}^{(n)}}\mathbf{u}_y \\
\mathbf{p}_\pm^{(n)} &= \mp\frac{k_z^{(n)}}{k_0 n_{hs}}\left(\frac{k_x^{(n)}}{k_{xy}^{(n)}}\mathbf{u}_x + \frac{k_y^{(0)}}{k_{xy}^{(n)}}\mathbf{u}_y\right) + \frac{k_{xy}^{(n)}}{k_0 n_{hs}}\mathbf{u}_z
\end{aligned}\right\} \tag{6.29}$$

denote linearly polarized fields of s– and p–types, in electromagnetics literature [45], with respect to the wavevector $\mathbf{k}_\pm^{(n)}$. The scalars

$$\left.\begin{aligned}
\kappa_x &= (\pi/\Omega)|\sin\alpha| \\
k_x^{(n)} &= k_x^{(0)} + n\kappa_x \\
k_z^{(n)} &= +\sqrt{k_0^2 n_{hs}^2 - \left(k_{xy}^{(n)}\right)^2} \\
k_{xy}^{(n)} &= +\sqrt{\left(k_x^{(n)}\right)^2 + \left(k_y^{(0)}\right)^2}
\end{aligned}\right\} \tag{6.30}$$

depend on the x–period $\Lambda_x = 2\Omega/|\sin\alpha|$ of the slanted chiral STF. The free–space wavenumber is denoted by $k_0 = \omega\sqrt{\mu_0\epsilon_0} = 2\pi/\lambda_0$.

The incident plane wave is the Floquet harmonic of order 0; hence, $a_L^{(n)} = a_R^{(n)} = 0 \,\forall n \neq 0$. Since all $\left\{a_L^{(n)}, a_R^{(n)}\right\}$ are supposed to be known, the amplitude sets $\left\{r_L^{(n)}, r_R^{(n)}\right\}$ and $\left\{t_L^{(n)}, t_R^{(n)}\right\}$ need to be determined by the solution of certain coupled–wave ordinary differential equations (ODEs) augmented by boundary conditions at $z = 0$ and $z = d$.

6.2.2.2 Coupled–Wave ODEs

The spatially periodic variation of $\underline{\epsilon}(\mathbf{r})$ of (6.16) is represented by the Fourier series

$$\underline{\epsilon}(\mathbf{r}) = \sum_{n\in\mathbb{Z}} \underline{\epsilon}^{(n)} \exp\left[in\left(\kappa_x x + \kappa_z z\right)\right], \quad 0 < z < d\,, \tag{6.31}$$

where

$$\underline{\epsilon}^{(n)} = \sum_{\sigma,\sigma'} \epsilon_{\sigma\sigma'}^{(n)} \mathbf{u}_\sigma \mathbf{u}_{\sigma'}, \quad \sigma, \sigma' = x, y, z, \quad (6.32)$$

are constant–value dyadics; and $\kappa_z = (\pi/\Omega)\cos\alpha$ is in accord with the z–period $\Lambda_z = 2\Omega/\cos\alpha$ of the slanted chiral STF.

Wave propagation occurs inside the thin–film material such that the electromagnetic field phasors everywhere can be decomposed as

$$\left. \begin{array}{l} \mathbf{E}(\mathbf{r}) = \sum_{n\in\mathbb{Z}} \left[E_x^{(n)}(z)\,\mathbf{u}_x + E_y^{(n)}(z)\,\mathbf{u}_y + E_z^{(n)}(z)\,\mathbf{u}_z \right] \exp\left[i\left(k_x^{(n)}x + k_y^{(0)}y\right) \right] \\[6pt] \mathbf{H}(\mathbf{r}) = \sum_{n\in\mathbb{Z}} \left[H_x^{(n)}(z)\,\mathbf{u}_x + H_y^{(n)}(z)\,\mathbf{u}_y + H_z^{(n)}(z)\,\mathbf{u}_z \right] \exp\left[i\left(k_x^{(n)}x + k_y^{(0)}y\right) \right] \end{array} \right\}, \quad (6.33)$$

where $E_{x,y,z}^{(n)}$ and $H_{x,y,z}^{(n)}$ are unknown functions of z.

On substituting (6.31)–(6.41) in the frequency–domain Maxwell curl postulates

$$\left. \begin{array}{l} \nabla \times \mathbf{E}(\mathbf{r}) = i\omega\mu_0\,\mathbf{H}(\mathbf{r}) \\[4pt] \nabla \times \mathbf{H}(\mathbf{r}) = -i\omega\epsilon_0\,\underline{\epsilon}(\mathbf{r}) \cdot \mathbf{E}(\mathbf{r}) \end{array} \right\}, \quad 0 < z < d, \quad (6.34)$$

and exploiting the orthogonalities of the functions $\exp(i\mathbf{k}_\pm^{(n)} \cdot \mathbf{r})$ across any plane $z = \text{constant}$, coupled–wave ODEs

$$\frac{d}{dz}\tilde{E}_x^{(n)}(z) + in\kappa_z\tilde{E}_x^{(n)}(z) - ik_x^{(n)}\tilde{E}_z^{(n)}(z) = ik_0\eta_0\tilde{H}_y^{(n)}(z), \quad (6.35)$$

$$\frac{d}{dz}\tilde{E}_y^{(n)}(z) + in\kappa_z\tilde{E}_y^{(n)}(z) - ik_y^{(0)}\tilde{E}_z^{(n)}(z) = -ik_0\eta_0\tilde{H}_x^{(n)}(z), \quad (6.36)$$

$$k_y^{(0)}\tilde{E}_x^{(n)}(z) - k_x^{(n)}\tilde{E}_y^{(n)}(z) = -k_0\eta_0\tilde{H}_z^{(n)}(z), \quad (6.37)$$

$$\frac{d}{dz}\tilde{H}_x^{(n)}(z) + in\kappa_z\tilde{H}_x^{(n)}(z) - ik_x^{(n)}\tilde{H}_z^{(n)}(z) = -\frac{ik_0}{\eta_0} \sum_{n'\in\mathbb{Z}} \left[\tilde{\epsilon}_{yx}^{(n-n')}\tilde{E}_x^{(n')}(z) \right.$$
$$\left. + \tilde{\epsilon}_{yy}^{(n-n')}\tilde{E}_y^{(n')}(z) + \tilde{\epsilon}_{yz}^{(n-n')}\tilde{E}_z^{(n')}(z) \right], \quad (6.38)$$

$$\frac{d}{dz}\tilde{H}_y^{(n)}(z) + in\kappa_z\tilde{H}_y^{(n)}(z) - ik_y^{(0)}\tilde{H}_z^{(n)}(z) = \frac{ik_0}{\eta_0} \sum_{n'\in\mathbb{Z}} \left[\tilde{\epsilon}_{xx}^{(n-n')}\tilde{E}_x^{(n')}(z) \right.$$
$$\left. + \tilde{\epsilon}_{xy}^{(n-n')}\tilde{E}_y^{(n')}(z) + \tilde{\epsilon}_{xz}^{(n-n')}\tilde{E}_z^{(n')}(z) \right], \quad (6.39)$$

$$k_y^{(0)}\tilde{H}_x^{(n)}(z) - k_x^{(n)}\tilde{H}_y^{(n)}(z) = \frac{k_0}{\eta_0} \sum_{n'\in\mathbb{Z}} \left[\tilde{\epsilon}_{zx}^{(n-n')}\tilde{E}_x^{(n')}(z) \right.$$
$$\left. + \tilde{\epsilon}_{zy}^{(n-n')}\tilde{E}_y^{(n')}(z) + \tilde{\epsilon}_{zz}^{(n-n')}\tilde{E}_z^{(n')}(z) \right] \quad (6.40)$$

are derived for $z \in (0, d)$, where

$$
\left.\begin{array}{l}
\tilde{E}_\sigma^{(n)}(z) = E_\sigma^{(n)}(z) \exp(-in\kappa_z z) \\
\tilde{H}_\sigma^{(n)}(z) = H_\sigma^{(n)}(z) \exp(-in\kappa_z z)
\end{array}\right\}, \quad \sigma = x, y, z. \tag{6.41}
$$

Equations (6.35)–(6.40) hold for all $n \in \mathbb{Z}$, and are thus an infinite system of first–order ODEs. For numerical solution, the restriction $|n| \le N_t$ is made. By substituting (6.37) and (6.40) into (6.35), (6.36), (6.38), and (6.39), in order to eliminate the normal electromagnetic field components (i.e., $\tilde{E}_z^{(n)}$ and $\tilde{H}_z^{(n)}$), and performing some algebraic manipulations, the first–order matrix ODE

$$
\frac{d}{dz}\left[\tilde{\underline{f}}(z)\right] = i\left[\underline{\underline{\tilde{P}}}\right]\left[\tilde{\underline{f}}(z)\right], \quad 0 < z < d, \tag{6.42}
$$

is eventually derived. The column vector

$$
\left[\tilde{\underline{f}}(z)\right] = \left[\left[\tilde{\underline{E}}_x(z)\right]^{\mathrm{T}}, \left[\tilde{\underline{E}}_y(z)\right]^{\mathrm{T}}, \eta_0\left[\tilde{\underline{H}}_x(z)\right]^{\mathrm{T}}, \eta_0\left[\tilde{\underline{H}}_y(z)\right]^{\mathrm{T}}\right]^{\mathrm{T}} \tag{6.43}
$$

contains $4(2N_t + 1)$ components, and the constant–valued matrix $\left[\underline{\underline{\tilde{P}}}\right]$ consists of sixteen $(2N_t + 1) \times (2N_t + 1)$ submatrices [51].

6.2.2.3 Solution of Boundary Value Problem

The matrix ODE (6.42) has the solution

$$
\left[\tilde{\underline{f}}(z_2)\right] = \left[\underline{\underline{\tilde{G}}}\right] \exp\left\{i(z_2 - z_1)\left[\underline{\underline{\tilde{D}}}\right]\right\} \left[\underline{\underline{\tilde{G}}}\right]^{-1} \left[\tilde{\underline{f}}(z_1)\right], \tag{6.44}
$$

where the columns of the square matrix $\left[\underline{\underline{\tilde{G}}}\right]$ are the successive eigenvectors of $\left[\underline{\underline{\tilde{P}}}\right]$, and the diagonal matrix $\left[\underline{\underline{\tilde{D}}}\right]$ contains the corresponding eigenvalues of $\left[\underline{\underline{\tilde{P}}}\right]$. The assumption here is that $\left[\underline{\underline{\tilde{P}}}\right]$ is diagonalizable, i.e., it has $4(2N_t + 1)$ linearly independent eigenvectors [66].

The relation

$$
[\underline{f}(z)] = \mathrm{Diag}\left[\exp\left(in_M\kappa_z z\right)\right]\left[\tilde{\underline{f}}(z)\right] \tag{6.45}
$$

yields the reformation of (6.44) as

$$
[\underline{f}(z_2)] = \left[\underline{\underline{G}}(z_2)\right] \exp\left\{i(z_2 - z_1)\left[\underline{\underline{\tilde{D}}}\right]\right\} \left[\underline{\underline{G}}(z_1)\right]^{-1} [\underline{f}(z_1)], \tag{6.46}
$$

where the matrix

$$
\left[\underline{\underline{G}}(z)\right] = \mathrm{Diag}\left[\exp\left(in_M\kappa_z z\right)\right]\left[\underline{\underline{\tilde{G}}}\right] \tag{6.47}
$$

is a periodic function of z, and $n_M = \text{Mod}[n - 1, 2N_t + 1] - N_t$. Hence,

$$[\mathbf{\underline{f}}(d)] = \left[\underline{\underline{G}}(d)\right] \exp\left\{id\left[\underline{\underline{\tilde{D}}}\right]\right\}\left[\underline{\underline{G}}(0)\right]^{-1}[\mathbf{\underline{f}}(0)] . \tag{6.48}$$

The continuity of the tangential components of the electric and magnetic field phasors across the two boundaries $z = 0$ and $z = d$ must be enforced with respect to the Floquet harmonic of any order n. From (6.20)–(6.25), the boundary values

$$[\mathbf{\underline{f}}(0)] = \begin{bmatrix} \left[\underline{\underline{Y}}_e^+\right] & \left[\underline{\underline{Y}}_e^-\right] \\ \left[\underline{\underline{Y}}_h^+\right] & \left[\underline{\underline{Y}}_h^-\right] \end{bmatrix}\begin{bmatrix} [A] \\ [R] \end{bmatrix}, \quad [\mathbf{\underline{f}}(d)] = \begin{bmatrix} \left[\underline{\underline{Y}}_e^+\right] & [0]_{4N_t+2} \\ \left[\underline{\underline{Y}}_h^+\right] & [0]_{4N_t+2} \end{bmatrix}\begin{bmatrix} [T] \\ [0]_{4N_t+2} \end{bmatrix}$$

$$\tag{6.49}$$

are available [51]. Here, $[A]$, $[R]$, and $[T]$ are $(4N_t + 2)$ column vectors comprising the amplitude sets $\left\{a_L^{(n)}, a_R^{(n)}\right\}$, $\left\{r_L^{(n)}, r_R^{(n)}\right\}$, and $\left\{t_L^{(n)}, t_R^{(n)}\right\}$, respectively.

Finally, the substitution of (6.49) into (6.48) yields

$$\begin{bmatrix} \left[\underline{\underline{U}}_T\right] \\ \left[\underline{\underline{V}}_T\right] \end{bmatrix}[T] + \begin{bmatrix} e^{id[\underline{\underline{\tilde{D}}}_u]} & [0]_{4N_t+2} \\ [0]_{4N_t+2} & e^{id[\underline{\underline{\tilde{D}}}_l]} \end{bmatrix}\begin{bmatrix} \left[\underline{\underline{U}}_R\right] \\ \left[\underline{\underline{V}}_R\right] \end{bmatrix}[R]$$

$$= \begin{bmatrix} e^{id[\underline{\underline{\tilde{D}}}_u]} & [0]_{4N_t+2} \\ [0]_{4N_t+2} & e^{id[\underline{\underline{\tilde{D}}}_l]} \end{bmatrix}\begin{bmatrix} \left[\underline{\underline{U}}_A\right] \\ \left[\underline{\underline{V}}_A\right] \end{bmatrix}[A] , \tag{6.50}$$

where $\left[\underline{\underline{\tilde{D}}}_u\right]$ and $\left[\underline{\underline{\tilde{D}}}_l\right]$ are the upper and lower diagonal submatrices of $\left[\underline{\underline{\tilde{D}}}\right]$, respectively.

For calculating the unknown $[R]$ and $[T]$, the so-called R–matrix propagating algorithm [67, 68] — which is based on the rearrangement of the positions of the eigenvalues of $\left[\underline{\underline{\tilde{P}}}\right]$ in the diagonal matrix $\left[\underline{\underline{\tilde{D}}}\right]$ — helps in avoiding numerical instabilities, especially when N_t is large [69, 70]. Therefore, the entries on the diagonal of $\left[\underline{\underline{\tilde{D}}}\right]$ (thus $\left[\underline{\underline{\tilde{D}}}_u\right]$ and $\left[\underline{\underline{\tilde{D}}}_l\right]$ also) are rearranged in the order of increasing magnitude of the imaginary part, and the columns of $\left[\underline{\underline{\tilde{G}}}\right]$ are rearranged accordingly. The final algebraic equation yielded by (6.50) for the determination of $[R]$ and $[T]$ is algorithmically stable due to the fact that the exponential terms $e^{-id[\underline{\underline{\tilde{D}}}_u]}$ and $e^{id[\underline{\underline{\tilde{D}}}_l]}$ never become overwhelming in magnitude because of the rearrangement of the eigenvalues.

6.2.2.4 Planewave Reflectance and Transmittance

Once the column vectors $[R]$ and $[T]$ have been determined from the solution of (6.50), the reflection and transmission coefficients

$$r_{\sigma\sigma'}^{(n)} = \frac{r_\sigma^{(n)}}{a_{\sigma'}^{(0)}}, \quad t_{\sigma\sigma'}^{(n)} = \frac{t_\sigma^{(n)}}{a_{\sigma'}^{(0)}}, \quad \sigma, \sigma' = L, R \tag{6.51}$$

can be computed as functions of the incidence wavevector $\mathbf{k}_+^{(0)}$. Reflectances $(R_{LL}^{(n)}$, etc.) and transmittances $(T_{LL}^{(n)}$, etc.) of order n are thereby defined as

$$R_{\sigma\sigma'}^{(n)} = \frac{\text{Re}[k_z^{(n)}]}{\text{Re}[k_z^{(0)}]} |r_{\sigma\sigma'}^{(n)}|^2, \quad T_{\sigma\sigma'}^{(n)} = \frac{\text{Re}[k_z^{(n)}]}{\text{Re}[k_z^{(0)}]} |t_{\sigma\sigma'}^{(n)}|^2, \quad \sigma, \sigma' = L, R, \quad (6.52)$$

where $\text{Re}[\ ]$ stands for the real part. Coefficients and remittances with both subscripts identical are co–polarized, and those with two different subscripts are cross–polarized. Co–polarized coefficients and remittances subscripted RR (LL) are labeled as co–handed, and those subscripted LL (RR) are labeled as cross–handed, when the chiral STF is structurally right (left)–handed.

Care must be taken for the special case $\alpha = 0$. All nonspecular coefficients then fold into the specular ones, i.e.,

$$r_{\sigma\sigma'}^{(0)} \Leftarrow \sum_{|n| \leq N_t} r_{\sigma\sigma'}^{(n)}, \quad t_{\sigma\sigma'}^{(0)} \Leftarrow \sum_{|n| \leq N_t} t_{\sigma\sigma'}^{(n)}, \quad \sigma, \sigma' = L, R, \quad (6.53)$$

because $\Lambda_x \to \infty$. Similarly, the remittances for $\alpha = 0$ can only be specular; hence,

$$R_{\sigma\sigma'}^{(0)} = |r_{\sigma\sigma'}^{(0)}|^2, \quad T_{\sigma\sigma'}^{(0)} = |t_{\sigma\sigma'}^{(0)}|^2, \quad \sigma, \sigma' = L, R. \quad (6.54)$$

6.2.3 Circular Bragg Phenomenon at Normal Incidence

6.2.3.1 Chiral STFs

To begin with, chiral STFs ($\alpha = 0$) must be examined. The response of this film to normally incident plane waves has been extensively studied, theoretically [9] as well as experimentally [5, 33, 71]. The hallmark of an axially excited chiral STF is the circular Bragg phenomenon: Provided the ratio $N_d = d/\Omega$ is sufficiently large and λ_0 lies within the Bragg regime, the reflectance is much higher if the handedness of the incident circularly polarized plane wave matches the structural handedness of the film than if otherwise. The center wavelength of the Bragg regime is estimated as [9]

$$\lambda_0^{Br} = \Omega \left[\sqrt{\epsilon_c \left(\lambda_0^{Br} \right)} + \sqrt{\epsilon_d \left(\lambda_0^{Br} \right)} \right]. \quad (6.55)$$

Coupled–wave theory provides an explanation for the polarization–sensitive Bragg phenomenon — a plane wave of matching handedness effectively encounters a Bragg grating, while that of the other handedness does not [30–32].

Figure 6.4 shows the reflectances and transmittances computed using the aforementioned numerical solution procedure versus those computed using the analytical procedure available elsewhere [9]. All reflectances and transmittances must be specular when $\alpha = 0$, regardless of their order n, in view of (6.53) and

(6.54). The Bragg regime — specified by $\lambda_0 \in (1050, 1150)$ nm — is obvious in the plots: $R_{RR}^{(0)} >> R_{LL}^{(0)}$ and $T_{RR}^{(0)} << T_{LL}^{(0)}$, while $R_{RL}^{(0)} \approx R_{LR}^{(0)}$ and $T_{RL}^{(0)} \approx T_{LR}^{(0)}$ are very small, especially around $\lambda_0^{Br} = 1090$ nm as predicted by (6.55). The coincidence of the two sets of computed remittances is evident in Figure 6.4.

6.2.3.2 Slanted Chiral STFs

Only for $\alpha \neq 0$ does a distinction between Floquet harmonics of order $n = 0$ and $n \neq 0$ appear, the former being classified as *specular* and the latter as *nonspecular* in the literature on diffraction gratings [45]. Nonnegligible remittances were found to be only of orders $n = \mp 2$ and $n = 0$.

Figure 6.5 comprises spectral plots of the reflectances and the transmittances of order $n = 0$ and -2 as functions of λ_0 when $\alpha = 15°$. The characteristic signature of the circular Bragg phenomenon is found as a broad crest in the plot of $R_{RR}^{(-2)}$ and a trough in the plot of $T_{RR}^{(0)}$, positioned between the 1030–nm and 1120–nm wavelengths. Thus, a normally incident RCP plane wave is mostly reflected obliquely: at an angle $\sin^{-1}[(\lambda_0/\Omega)\sin\alpha]$ to the $+z$ axis.

When the incident plane wave is LCP, however, $T_{LL}^{(0)}$ dominates all the other remittances in the same wavelength regime. Hence, the LCP plane wave is mostly transmitted without change of direction — in a fashion similar to that for $\alpha = 0$.

Clearly, the circular Bragg phenomenon is partly nonspecular for slanted chiral STFs, as indicated by a high co–handed reflectance of order $n = \mp 2$ for $\alpha \gtrless 0$ and a low co–handed specular transmittance in the Bragg regime. This characteristic of the circular Bragg phenomenon can be exploited for circular–polarization beamsplitters and direction couplers.

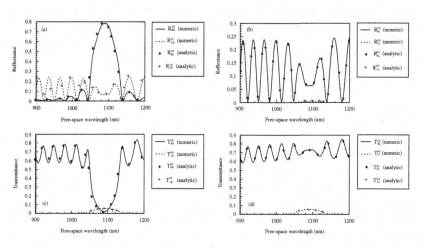

FIGURE 6.4. Reflectances and transmittances computed using the numerical solution procedure of Section 6.2 versus the results obtained from the analytical method available elsewhere [9]. The various parameters used are as follows: $p_a = 2.0$, $p_b = 2.6$, $p_c = 2.1$, $N_a = N_b = N_c = 100$, $\lambda_a = \lambda_c = 140$ nm, $\lambda_b = 150$ nm, $\Omega = 300$ nm, $d = 27\Omega$, $\alpha = 0$, $\chi_s = 30°$, $h = 1$, $n_{hs} = 1$, and $\theta_i^p = \psi_i^p = 0$.

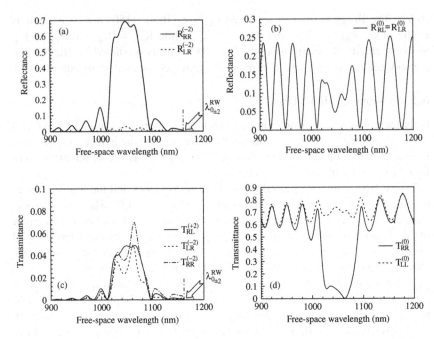

FIGURE 6.5. Reflectances and transmittances of Floquet–harmonic order n, computed for the same parameters as for Figure 6.4 but with $\alpha = 15°$. Reflectances and transmittances of maximum magnitudes less than 0.01 are not shown.

The center wavelength of the Bragg regime shifts to smaller values with increasing $|\alpha|$. This blue shift is captured by the modification of (6.55) to

$$\lambda_0^{Br} = (\Omega \cos\alpha) \left[\sqrt{\epsilon_c \left(\lambda_0^{Br}\right)} + \sqrt{\epsilon_d \left(\lambda_0^{Br}\right)} \right] \tag{6.56}$$

for $|\alpha| << \chi_s$, as suggested by the values of λ_0^{Br} computed using the aforementioned numerical solution procedure for various $|\alpha|$.

Figure 6.6a shows a comparison of the $|\alpha|$–dependence of λ_0^{Br} obtained numerically with (6.56), and the excellent correspondence for $|\alpha| \leq 15°$ is clear. Not surprisingly, this $|\alpha|$–dependence of λ_0^{Br} as predicted by (6.56) is in accord with the orange color of multidomain cholesteric liquid crystals that were fabricated to reflect the red color [72].

Figure 6.6b contains a plot of the full–width–at–half–maximum (FWHM) bandwidth $\Delta\lambda_0^{Br}$ as a function of $|\alpha|$. This plot clearly shows the thinning of the Bragg regime with increase in $|\alpha|$. Indeed, the circular Bragg phenomenon vanishes for $|\alpha| > 17.1°$. The reason for the disappearance of that phenomenon is best explained by the so–called Rayleigh–Wood anomalies as follows.

Just as for the commonplace surface–relief gratings [45], nonspecular Floquet harmonics exist in the two half–spaces $z \leq 0$ and $z \geq d$. A Floquet harmonic of

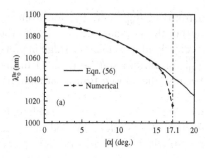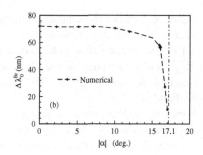

FIGURE 6.6. (a) Center wavelength λ_0^{Br} of the Bragg regime as a function of $|\alpha|$, computed using (6.56) (solid line) and the numerical solution procedure of Section 6.2 (square–dashed line). (b) The FWHM bandwidth $\Delta\lambda_0^{Br}$ of the Bragg regime computed using the numerical solution procedure of Section 6.2.

order n propagates energy away from the slanted chiral STF, provided $k_z^{(n)}$ is real–valued, i.e., for wavelengths $\lambda_0 < \lambda_{0_n}^{RW}$, where the relation

$$\lambda_{0_n}^{RW} = \frac{2n_{hs}\Omega}{|n|\,|\sin\alpha|} \tag{6.57}$$

follows from (6.30). The conversion of the Floquet harmonic of order n from propagating to evanescent, or vice versa, as λ_0 either increases or decreases across $\lambda_{0_n}^{RW}$ is a Rayleigh–Wood anomaly.

As $|\alpha|$ increases from 0, $\lambda_{0_{\mp2}}^{RW}$ decreases from "infinity" and begins to approach the center wavelength λ_0^{Br} of the Bragg regime predicted by (6.56). For example, when $|\alpha| = 15°$, $\lambda_{0_{\mp2}}^{RW}$ is about 100 nm larger than λ_0^{Br}. Still, the spectrums of $R_{RR}^{(\mp2)}$ and $T_{RR}^{(0)}$ are clearly affected, as implied by Figure 6.5. A further increase in $|\alpha|$ makes $\lambda_{0_{\mp2}}^{RW}$ come even closer to λ_0^{Br}, leading to the shrinkage of the Bragg regime and the subversion of the circular Bragg phenomenon to a considerable level. Eventually, the circular Bragg phenomenon disappears as a result of the increase of $|\alpha|$ to make the nonspecular parts become negligible.

6.2.4 Circular Bragg Phenomenon at Oblique Incidence

In general, the circular Bragg phenomenon exists also for oblique incidence of plane waves, but it is greatly influenced by the directionality of planewave incidence, as is known for $\alpha = 0$ [27, 28]. Therefore, the spectral (i.e., wavelength–dependence) and angular–spread (i.e., wavevector–dependence) features of the circular Bragg phenomenon are interwound for the case of oblique incidence [51]. [1]

---

[1] Let the wavevector $\mathbf{k}_+^{(0)}$ make the angles (i) $\theta_i^p \in (-90°, 90°)$ to the z axis and (ii) $\psi_i^p \in [0, 180°)$ to the x axis in the xy plane.

Figure 6.7 shows the characteristic spectrums of $R_{RR}^{(-2)}$ and $T_{RR}^{(0)}$ for arbitrary incidence in either the xz plane (i.e., $\psi_i^p = 0$) or the yz plane (i.e., $\psi_i^p = 90°$) for $\alpha = 10°$. The circular Bragg phenomenon is clearly identifiable as a ridge in plots of $R_{RR}^{(-2)}$ and as a trough in the plots of $T_{RR}^{(0)}$ for various $\sin\theta_i^p \in (-1, 1)$ in the wavelength regime $\lambda_0 \in [650, 750]$ nm, because the chosen film is structurally right–handed.

Figure 6.7 indicates that the Bragg regime blue–shifts more for more obliquely incident plane waves. Moreover, the influence of a Rayleigh–Wood anomaly on the circular Bragg phenomenon is very explicit for obliquely incident plane waves. The angular spread of the Bragg regime — quantitated as the θ_i^p span of the broad ridge in the plot of $R_{RR}^{(-2)}$ — becomes asymmetric with respect to θ_i^p in the xz plane (Figure 6.7a), but not in the yz plane (Figure 6.7c).

In fact, because of the incidence being oblique, both λ_0^{Br} and $\lambda_{0,n}^{RW}$ depend on the values of θ_i^p and ψ_i^p. Therefore, (6.56) and (6.57) must undergo changes for oblique incidence. On the one hand, computations suggest that

$$\lambda_0^{Br} = \Omega \cos\alpha \sqrt{\cos\theta_i^p} \left[\sqrt{\epsilon_c(\lambda_0^{Br})} + \sqrt{\epsilon_d(\lambda_0^{Br})} \right] \tag{6.58}$$

provides a good quantitative estimate of λ_0^{Br} when $|\theta_i^p| \leq 30°$, although the actual functional dependence of λ_0^{Br} on both θ_i^p and ψ_i^p might be very complicated. On the other hand, the Rayleigh–Wood anomalies of different orders occur at

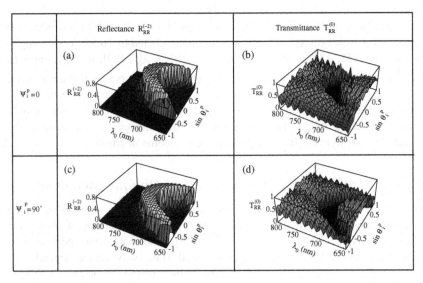

FIGURE 6.7. Spectrums of reflectance $R_{RR}^{(-2)}$ and transmittance $T_{RR}^{(0)}$, computed for a slanted chiral STF for oblique planewave incidence. (a, b) $\psi_i^p = 0$; (c, d) $\psi_i^p = 90°$. The various parameters used are as follows: $p_a = 2.0$, $p_b = 2.6$, $p_c = 2.1$, $N_a = N_b = N_c = 500$, $\lambda_a = \lambda_c = 140$ nm, $\lambda_b = 150$ nm, $\Omega = 200$ nm, $d = 60\Omega$, $\alpha = 10°$, $\chi_s = 30°$, $h = 1$, and $n_{hs} = 1$.

$$\lambda_{0_n}^{\mathrm{RW}} = \frac{n_{hs}\Omega}{|n\sin\alpha|}\left[\sqrt{1-\left(\sin\theta_i^p\sin\psi_i^p\right)^2}-\sin\theta_i^p\cos\psi_i^p\,\mathrm{sign}(n\alpha)\right] \qquad (6.59)$$

for oblique incidence in general. Clearly, $\lambda_{0_n}^{\mathrm{RW}}\neq\lambda_{0_{-n}}^{\mathrm{RW}}$ when $\theta_i^p\neq 0$ and $\psi_i^p\neq\pm 90°$.

As the wavevector $\mathbf{k}_+^{(0)}$ tilts away from the z axis, both λ_+^{Br} and $\lambda_{0_n}^{\mathrm{RW}}$ (for $\alpha\gtrless 0$) change noticeably. If $\lambda_{0_{\mp 2}}^{\mathrm{RW}}$ becomes smaller than λ_0^{Br}, the signature of the circular Bragg phenomenon disappears. The wavelength neighborhood of the disappearance depends strongly on the orientation of the plane of incidence with respect to the plane containing the helical axis (i.e., xz plane); hence, the circular Bragg phenomenon is far from displaying circular symmetry with respect to ψ_i^p when $\alpha\neq 0$, as seen from the contrast between Figures 6.7a and 6.7c.

The dependence of $\lambda_{0_{\mp 2}}^{\mathrm{RW}}$ (for $\alpha\gtrless 0$) on $\sin\theta_i^p\cos\psi_i^p$ is different from that on $\sin\theta_i^p\sin\psi_i^p$. When $\sin\theta_i^p\sin\psi_i^p$ is constant–valued, $\lambda_{0_{\mp 2}}^{\mathrm{RW}}$ changes linearly with $\sin\theta_i^p\cos\psi_i^p$; and the Bragg regime is susceptible to subversion by the Rayleigh–Wood anomaly of order $n=\mp 2$ if $\sin\theta_i^p\cos\psi_i^p\lessgtr 0$. That is the reason for the absence of the Bragg regime in Figures 6.7a and 6.7b ($\psi_i^p=0$), for $\sin\theta_i^p\in(-1,-0.37)$. In contrast, $\lambda_{0_{\mp 2}}^{\mathrm{RW}}$ is a monotonically decreasing function of $|\sin\theta_i^p\sin\psi_i^p|$ for constant–valued $\sin\theta_i^p\cos\psi_i^p$. Therefore, in Figures 6.7c and 6.7d ($\psi_i^p=90°$), the Bragg regime is completely subverted by the Rayleigh–Wood anomaly of order $n=-2$ in the angular regime $|\sin\theta_i^p|\in(0.83,1)$.

6.3 Spectral Holes in Slanted Chiral STFs

Previous studies have shown the emergence of spectral holes in the Bragg regime of two–section chiral STFs containing either layer or twist defects or both [15, 37, 58]. There are two types of spectral holes in a chiral STF with a central defect, and the evolution of these spectral holes with the thickness of the chiral STF sections gives evidence of a remarkable crossover phenomenon [62, 63]. The objective of this section is to identify the spectral holes and the associated crossover phenomenon in slanted chiral STFs with central twist defects.

6.3.1 Geometry of Twist Defect

The geometry of a slanted chiral STF with a central twist defect is sketched in Figure 6.8. The slanted chiral STF has a thickness of d and is bounded by two vacuous half–spaces (i.e., $n_{hs}=1$). The central twist defect is introduced by a twist angle $\phi\neq m\pi$ ($m\in\mathbb{Z}$) between the upper and the lower halves about their common axis of nonhomogeneity. The constitutive relations of the slanted chiral STF are delineated in the same way as presented in (6.15) and (6.16), except that the rotational dyadic $\underline{\underline{s}}_z(\mathbf{r})$ of (6.16) is reformulated as

$$\underline{\underline{s}}_z(\mathbf{r}) = \left(\mathbf{u}_x\mathbf{u}_x+\mathbf{u}_y\mathbf{u}_y\right)\cos\left[\frac{\pi}{\Omega}\left(\mathbf{r}\cdot\mathbf{u}_\ell\right)+\phi_{layer}\right]$$
$$+h\left(\mathbf{u}_y\mathbf{u}_x-\mathbf{u}_x\mathbf{u}_y\right)\sin\left[\frac{\pi}{\Omega}\left(\mathbf{r}\cdot\mathbf{u}_\ell\right)+\phi_{layer}\right]+\mathbf{u}_z\mathbf{u}_z\,, \qquad (6.60)$$

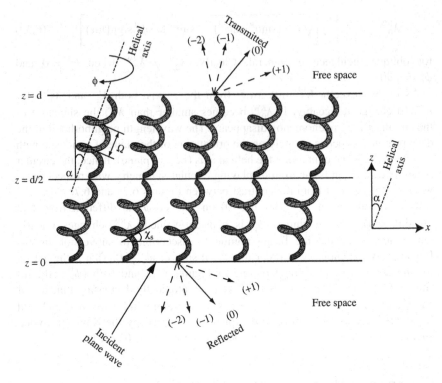

FIGURE 6.8. Schematic of the boundary value problem involving a slanted chiral STF with a central twist defect. The twist defect is introduced by a twist angle $\phi \neq m\pi$ $(m \in \mathbb{Z})$ between the upper and the lower halves about the axis of nonhomogeneity. Nonspecular reflection and transmission can occur because $\alpha \neq 0$.

where the auxiliary angle

$$\phi_{layer} = \begin{cases} 0, & 0 < z < d/2, \\ \phi, & d/2 < z < d. \end{cases} \tag{6.61}$$

6.3.2 Planewave Solution Procedure

The slanted chiral STF with a central twist defect can be viewed as a two–section slanted chiral STF, each section of which has a different permittivity dyadic. A numerical solution procedure is devised in Section 6.2 for the planewave response of a single–section slanted chiral STF. Therefore, a similar solution procedure can be devised for the planewave response of a slanted chiral STF with a central twist defect.

After following the exact steps from (6.20) to (6.42), the first–order matrix ODE

$$\frac{d}{dz}\left[\tilde{\mathbf{f}}(z)\right] = i\left[\underline{\tilde{\mathbf{P}}}(z)\right]\left[\tilde{\mathbf{f}}(z)\right], \quad 0 < z < d, \tag{6.62}$$

is derived for the slanted chiral STF being considered. In the presence of a central twist defect, the matrix $\left[\tilde{\underline{\underline{P}}}(z)\right]$ becomes piecewise uniform such that

$$\left[\tilde{\underline{\underline{P}}}(z)\right] = \begin{cases} \left[\tilde{\underline{\underline{P}}}\right]_1, & 0 < z < d/2, \\ \left[\tilde{\underline{\underline{P}}}\right]_2, & d/2 < z < d, \end{cases} \tag{6.63}$$

where $\left[\tilde{\underline{\underline{P}}}\right]_\sigma$, $(\sigma = 1, 2)$, is a constant–valued matrix. In accord with (6.61), it is evident that $\left[\tilde{\underline{\underline{P}}}\right]_1 = \left[\tilde{\underline{\underline{P}}}\right]$ and $\left[\tilde{\underline{\underline{P}}}\right]_2$ is related to $\left[\tilde{\underline{\underline{P}}}\right]$ with the replacement

$$\left[\underline{\underline{\epsilon}}_{\sigma,\sigma'}\right] \Rightarrow \mathrm{Diag}[e^{in\phi}]_{[-N_t,N_t]} \left[\underline{\underline{\epsilon}}_{\sigma,\sigma'}\right] \mathrm{Diag}[e^{-in\phi}]_{[-N_t,N_t]} \tag{6.64}$$

to derive its expression.

To be concise, let $d_\sigma = \sigma d/2$, $(\sigma = 0, 1, 2)$. According to (6.44), the matrix ODE (6.62) has the solution

$$[\underline{f}(d_{\sigma-1})] = \left[\tilde{\underline{\underline{G}}}\right]_\sigma \exp\left\{\frac{-id}{2}\left[\tilde{\underline{\underline{D}}}\right]_\sigma\right\}\left[\tilde{\underline{\underline{G}}}\right]_\sigma^{-1} [\underline{f}(d_\sigma)], \quad \sigma = 1, 2, \tag{6.65}$$

where the square matrix $\left[\tilde{\underline{\underline{G}}}\right]_\sigma$ consists of the eigenvectors of $\left[\tilde{\underline{\underline{P}}}\right]_\sigma$ as its columns; and the diagonal matrix $\left[\tilde{\underline{\underline{D}}}\right]_\sigma$ contains the eigenvalues of $\left[\tilde{\underline{\underline{P}}}\right]_\sigma$ in the same order. Combination of (6.45) and (6.65) yields

$$[\underline{f}(d_{\sigma-1})] = \left[\underline{\underline{G}}(d_{\sigma-1})\right]_\sigma \exp\left\{\frac{-id}{2}\left[\tilde{\underline{\underline{D}}}\right]_\sigma\right\}\left[\underline{\underline{G}}(d_\sigma)\right]_\sigma^{-1} [\underline{f}(d_\sigma)], \quad \sigma = 1, 2,$$
$$\tag{6.66}$$

where the matrix

$$\left[\underline{\underline{G}}(z)\right]_\sigma = \mathrm{Diag}\left[\exp\left(in_M \kappa_z z\right)\right]\left[\tilde{\underline{\underline{G}}}\right]_\sigma. \tag{6.67}$$

Augmented by the boundary values (6.49), the iterative relation (6.66) suffices to determine the unknown $[\underline{R}]$ and $[\underline{T}]$. To avoid numerical instability, it is necessary to formulate a stable algorithm from (6.66). This can be done as follows [69, 70, 73]: Let

$$[\underline{f}(d_\sigma)] = \left[\underline{\underline{T}}\right]_\sigma [\underline{T}]_\sigma, \quad \sigma = 0, 1, 2, \tag{6.68}$$

where

$$[\underline{T}]_2 = [\underline{T}], \quad \left[\underline{\underline{T}}\right]_2 = \begin{bmatrix} \left[\underline{Y}_e^+\right] \\ \left[\underline{Y}_h^+\right] \end{bmatrix}, \tag{6.69}$$

and

$$[\underline{T}]_{\sigma-1} = e^{\frac{-id}{2}[\underline{\tilde{D}}_u]_\sigma} [\underline{\underline{v}}_u]_\sigma [\underline{T}]_\sigma , \quad \sigma = 1, 2. \tag{6.70}$$

Here, $[\underline{\tilde{D}}_u]_\sigma$ (and $[\underline{\tilde{D}}_l]_\sigma$) is the upper (lower) diagonal submatrices of $[\underline{\tilde{D}}]_\sigma$, while $[\underline{v}_u]_\sigma$ and $[\underline{v}_l]_\sigma$ are defined through the relation

$$\left[\begin{matrix} [\underline{v}_u]_\sigma \\ [\underline{v}_l]_\sigma \end{matrix} \right] = [\underline{G}(d_\sigma)]_\sigma^{-1} [\underline{T}]_\sigma , \quad \sigma = 1, 2. \tag{6.71}$$

The substitution of (6.68) and (6.71) into (6.65) yields

$$[\underline{T}]_{\sigma-1} = [\underline{G}(d_{\sigma-1})]_\sigma \left[\begin{matrix} [\underline{I}]_{4N_t+2} \\ e^{\frac{-id}{2}[\underline{\tilde{D}}_l]_\sigma} [\underline{v}_l]_\sigma \{[\underline{v}_u]_\sigma\}^{-1} e^{\frac{id}{2}[\underline{\tilde{D}}_u]_\sigma} \end{matrix} \right], \tag{6.72}$$

which, combined with (6.71), establishes the iterative relation for $[\underline{T}]_\sigma$. Here, $[\underline{I}]_m$ denotes the $m \times m$ identity matrix. Algorithm stability is guaranteed for (6.72) by rearrangement of eigenvalues on the diagonal of $[\underline{\tilde{D}}]_\sigma$ in the order of decreasing magnitude of the imaginary part.

From (6.70)–(6.72), the expressions of $[\underline{T}]_0$ and $[\underline{T}]_0$ are obtained in terms of $[\underline{T}]_2$ and $[\underline{T}]_2$ of (6.69). After partitioning

$$[\underline{T}]_0 = \left[\begin{matrix} [\underline{T}_u]_0 \\ [\underline{T}_l]_0 \end{matrix} \right], \tag{6.73}$$

and on account of (6.49) and (6.68), $[\underline{R}]$ and $[\underline{T}]_0$ are determined as

$$\left[\begin{matrix} [\underline{T}]_0 \\ [\underline{R}] \end{matrix} \right] = \left[\begin{matrix} [\underline{T}_u]_0 & -[\underline{Y}_e^-] \\ [\underline{T}_l]_0 & -[\underline{Y}_h^-] \end{matrix} \right]^{-1} \left[\begin{matrix} [\underline{Y}_e^+] \\ [\underline{Y}_h^+] \end{matrix} \right] [\underline{A}] . \tag{6.74}$$

Once $[\underline{T}]_0$ has been determined, $[\underline{T}] = [\underline{T}]_2$ is easily obtained by reverse iteration of (6.70).

6.3.3 Crossover Phenomenon of Spectral Holes

Let us consider the optical response of a slanted chiral STF with a central 90° twist defect. Due to the twist defect, a spectral hole is produced in the center of the Bragg regime, as the preceding studies had suggested for the chiral STFs [37, 58]. Figure 6.9 shows a narrow co–handed reflectance hole (of about 2–nm bandwidth) in $R_{RR}^{(-2)}$ and correspondingly, a co–handed transmittance peak in the $T_{RR}^{(0)}$ at the peak wavelength $\lambda_0^p = \lambda_0^{Br} = 1053$ nm, when $\alpha = 15°$ and the thickness ratio $N_d = 54$. Since the circular Bragg phenomenon is partly nonspecular for $\alpha \neq 0$, the co–handed reflectance hole is nonspecular for $\alpha \neq 0$.

A cross–handed spectral hole must also be excitable for the slanted chiral STF, provided the thickness ratio N_d is sufficiently large. Indeed, Figure 6.10 shows an ultranarrow cross–handed transmittance hole (about 0.15–nm bandwidth) in $T_{LL}^{(0)}$, and it is accompanied by the cross–handed transmittance/reflectance peaks in $R_{LL}^{(-2)}$, $R_{LL}^{(0)}$, and $T_{LL}^{(-2)}$ at the peak wavelength $\lambda_0^p = 1052.80$ nm, when $\alpha = 15°$ and $N_d = 182$. Clearly, the cross–handed transmittance is in the specular mode.

Therefore, there are two types of spectral holes of opposite circular polarization states, depending on different ranges of the value of N_d. As the two types of spectral holes evolve with the increase of N_d, a crossover phenomenon can be identified, and the crossover value of N_d, denoted by N_d^{co}, can be defined consequently to differentiate the two N_d ranges. Typically, the bandwidth of the cross–handed transmittance hole is extraordinarily small, and is independent of N_d or saturated beyond its crossover value N_d^{co}.

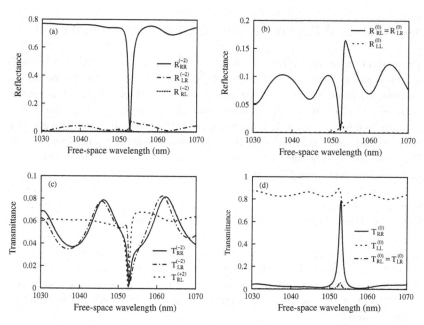

FIGURE 6.9. (a, b) Nonspecular and specular reflectances and (c, d) nonspecular and specular transmittances of order n, computed for the chiral STF with the following parameters: $p_a = 2.0$, $p_b = 2.6$, $p_c = 2.1$, $N_a = N_b = N_c = 4000$, $\lambda_a = \lambda_c = 140$ nm, $\lambda_b = 150$ nm, $\Omega = 300$ nm, $d = 54\Omega$, $\alpha = 15°$, $\chi_s = 30°$, $\phi = 90°$, $h = 1$, $n_{hs} = 1$, and $\theta_i^p = \psi_i^p = 0$. Remittances of maximum magnitudes less than 0.01 are not shown.

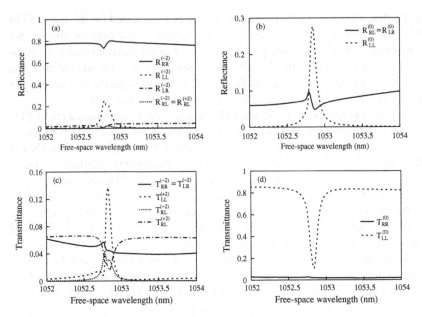

FIGURE 6.10. Same as Figure 6.9, but for $d = 182\Omega$.

6.4 Analytical Reconstruction of Crossover Phenomenon

The exhibition of two types of spectral holes in a slanted chiral STF with a central twist defect is established in Section 6.3, based on the numerical solution procedure devised in Sections 6.2 and 6.3. Notably, there is a crossover phenomenon associated with the occurrence of the two types of spectral holes: A co–handed reflectance hole is seen when the chosen thin film is relatively thin, but a cross–handed transmittance hole takes over when the thickness is large. The same crossover phenomenon also occurs in other types of chiral structures, such as chiral STFs and CLCs [62, 63].

There is a need to theoretically understand the crossover phenomenon rather than simply examine it numerically. Kopp and Genack [59] and Schmidtke and Stille [61] formulated the existence of a localized defect mode to explain the crossover phenomenon in CLCs. Oldano [74] and Becchi et al. [60] pointed out that the defect mode in CLCs has both spatially localized and nonlocalized components, because the electromagnetic modes in an axially excited CLC or chiral STF are not circularly polarized, in general [75]. Both Oldano and Kopp–Genack agreed that an analytic explanation of the crossover phenomenon is greatly desirable.

The objective of this section is to present an analytical reconstruction and explanation of the crossover phenomenon for chiral STFs. A similar explanation should hold for slanted chiral STFs as well, but the focus is kept on $\alpha = 0$ to keep analysis tractable. An analytic treatment of the central 90° twist defect in a chiral STF is presented here.

6.4.1 Coupled–Wave Theory

6.4.1.1 Coupled–Wave ODEs

Let the region $0 < z < d$ be occupied by a chiral STF with a central twist defect of $\phi = 90°$, while the two half–spaces $z \leq 0$ and $z \geq d$ are filled with a homogeneous, isotropic, dielectric medium of refractive index n_{hs}. A plane wave is normally incident onto the plane $z = 0$ from the half–space $z \leq 0$. As a result, the chiral STF is axially excited, such that the electromagnetic field phasors $\mathbf{E}(\mathbf{r})$ and $\mathbf{H}(\mathbf{r})$ are independent of x and y. The Maxwell curl postulates in (6.34) now reduce to the ODEs

$$\left.\begin{array}{l} \mathbf{u}_z \times \frac{d}{dz}\mathbf{E}(z) = i\omega\mu_0\mathbf{H}(z) \\[2mm] \mathbf{u}_z \times \frac{d}{dz}\mathbf{H}(z) = -i\omega\epsilon_0\underline{\underline{\epsilon}}(z)\cdot\mathbf{E}(z) \end{array}\right\}, \quad 0 < z < d, \tag{6.75}$$

for axial wave propagation in the chiral STF. By projecting (6.75) onto the subspace perpendicular to the z axis (i.e., xy plane), the following matrix ODEs for axial wave propagation are obtained [32]:

$$\left.\begin{array}{l} \frac{d}{dz}\left(\begin{bmatrix} 0 & -1 \\ 1 & 0 \end{bmatrix}[\underline{E}_\perp(z)]\right) = ik_0\,[\underline{H}'_\perp(z)] \\[4mm] \frac{d}{dz}\left(\begin{bmatrix} 0 & -1 \\ 1 & 0 \end{bmatrix}[\underline{H}'_\perp(z)]\right) = -ik_0\left[\underline{\underline{\epsilon}}_{perp}(z)\right][\underline{E}_\perp(z)] \end{array}\right\}, \quad 0 < z < d. \tag{6.76}$$

Here,

$$[\underline{E}_\perp(z)] = \begin{bmatrix} E_x(z), & E_y(z) \end{bmatrix}^{\mathrm{T}}, \quad [\underline{H}'_\perp(z)] = \begin{bmatrix} \eta_0 H_x(z), & \eta_0 H_y(z) \end{bmatrix}^{\mathrm{T}} \tag{6.77}$$

are 2–column vectors;

$$\left[\underline{\underline{\epsilon}}_{perp}(z)\right] = \left\{\left[\underline{\underline{\epsilon}}(z)\right]^{-1}_\perp\right\}^{-1} \tag{6.78}$$

is a 2×2 matrix; $\left[\underline{\underline{\epsilon}}(z)\right]$ is the matrix form of $\underline{\underline{\epsilon}}(z)$; and $\left[\underline{\underline{\epsilon}}(z)\right]^{-1}_\perp$ is obtained by neglecting all components of $\left[\underline{\underline{\epsilon}}(z)\right]^{-1}$ on the z axis.

By simply eliminating $[\underline{H}'_\perp(z)]$ in (6.76), the matrix ODE

$$\frac{d^2}{dz^2}[\underline{E}_\perp(z)] + k_0^2\left[\underline{\underline{\epsilon}}_{perp}(z)\right][\underline{E}_\perp(z)] = [\underline{0}]_2\,, \tag{6.79}$$

analogous to the homogeneous Helmholtz equation, is obtained for the axial variation of $[\underline{E}_\perp(z)]$.

Analytic solution of the second–order ODE (6.79) needs an appropriate expansion of the matrix $\left[\underline{\underline{\epsilon}}_{perp}(z)\right]$. The axial periodicity of chiral STFs suggests that $\left[\underline{\underline{\epsilon}}_{perp}(z)\right]$ can be decomposed into a Fourier series [76]. According to (6.16) and (6.61), the Fourier representation of $\left[\underline{\underline{\epsilon}}_{perp}(z)\right]$ in (6.78) is

$$\left[\underline{\underline{\epsilon}}_{perp}(z)\right] = \begin{cases} \bar{\epsilon}\left[\underline{\underline{I}}\right]_2 + \delta_\epsilon\left[\underline{\underline{F}}\right]e^{i2\pi z/\Omega} + \delta_\epsilon\left[\underline{\underline{F}}\right]^* e^{-i2\pi z/\Omega}, & 0 < z < d, \\ \bar{\epsilon}\left[\underline{\underline{I}}\right]_2 - \delta_\epsilon\left[\underline{\underline{F}}\right]e^{i2\pi z/\Omega} - \delta_\epsilon\left[\underline{\underline{F}}\right]^* e^{-i2\pi z/\Omega}, & d < z < 2d, \end{cases}$$

(6.80)

where

$$\left.\begin{matrix} \bar{\epsilon} = (\epsilon_d + \epsilon_c)/2 \\ \delta_\epsilon = (\epsilon_d - \epsilon_c)/2 \end{matrix}\right\}, \quad \left[\underline{\underline{F}}\right] = \begin{bmatrix} 1 & -hi \\ -hi & -1 \end{bmatrix},$$

(6.81)

and the superscript $*$ denotes the complex conjugate.

Naturally, the ODE (6.79) yields the solution

$$[E_\perp(z)] = \left[\underline{B}^+(z)\right]e^{ikz} + \left[\underline{B}^-(z)\right]e^{-ikz},$$

(6.82)

where $k = k_0\bar{n}$, $\bar{n} = \sqrt{\bar{\epsilon}}$, and $\left[\underline{\breve{B}}^\pm(z)\right]$ are 2–column vectors with the z–dependence accounting for the periodic nature of $\left[\underline{\underline{\epsilon}}_{perp}(z)\right]$.

Furthermore, in light of the circular–polarization–discriminatory optical responses of chiral STFs, it is useful to transform $\left[\underline{B}^\pm(z)\right]$ into their CP counterparts $\left[\underline{\breve{B}}^\pm(z)\right]$ by

$$\left[\underline{B}^\pm(z)\right] = \left[\underline{\underline{Y}}\right]\left[\underline{\breve{B}}^\pm(z)\right],$$

(6.83)

where the transformation matrix

$$\left[\underline{\underline{Y}}\right] = \frac{1}{\sqrt{2}}\begin{bmatrix} 1 & 1 \\ i & -i \end{bmatrix},$$

(6.84)

and the components of the 2–column vectors

$$\left[\underline{\breve{B}}^+(z)\right] = \begin{bmatrix} B_L^+(z) \\ B_R^+(z) \end{bmatrix}, \quad \left[\underline{\breve{B}}^-(z)\right] = \begin{bmatrix} B_R^-(z) \\ B_L^-(z) \end{bmatrix}$$

(6.85)

are LCP and RCP, as indicated by the subscripts L and R.

By substituting (6.80), (6.82), and (6.83) into the ODE (6.79), and enforcing the mutual orthogonalities of LCP and RCP phasors, the following coupled–wave ODEs are derived:

$$\left.\begin{aligned} \frac{1}{k_0^2}\frac{d^2}{dz^2}\left[\underline{\breve{B}}^+(z)\right] + \frac{2ik}{k_0^2}\frac{d}{dz}\left[\underline{\breve{B}}^+(z)\right] &= \mp\delta_\epsilon\left\{\left[\underline{\underline{\tilde{F}}}^+\right]\left[\underline{\breve{B}}^-(z)\right]e^{-2i(k-\pi/\Omega)z}\right. \\ &\qquad + \left[\underline{\underline{\tilde{F}}}^-\right]\left[\underline{\breve{B}}^-(z)\right]e^{-2i(k+\pi/\Omega)z}\Big\} \\ \frac{1}{k_0^2}\frac{d^2}{dz^2}\left[\underline{\breve{B}}^-(z)\right] - \frac{2ik}{k_0^2}\frac{d}{dz}\left[\underline{\breve{B}}^-(z)\right] &= \mp\delta_\epsilon\left\{\left[\underline{\underline{\tilde{F}}}^-\right]\left[\underline{\breve{B}}^+(z)\right]e^{2i(k-\pi/\Omega)z}\right. \\ &\qquad + \left[\underline{\underline{\tilde{F}}}^+\right]\left[\underline{\breve{B}}^+(z)\right]e^{2i(k+\pi/\Omega)z}\Big\} \end{aligned}\right\}.$$

(6.86)

In these equations, the upper signs (before δ_ϵ) apply for $0 < z < d/2$ and the lower signs for $d/2 < z < d$, while the 2×2 matrices

$$\left[\underline{\underline{F}}^{\pm}\right] = \frac{1}{2}\begin{bmatrix} 0 & 1 \mp h \\ 1 \pm h & 0 \end{bmatrix} \tag{6.87}$$

are antidiagonal. Hence, the ODEs (6.86) decouple $B_L^{\pm}(z)$ from $B_R^{\pm}(z)$, but $B_L^-(z)$ is coupled to $B_L^+(z)$, and $B_R^-(z)$ to $B_R^+(z)$.

6.4.1.2 Transfer Matrices

After substituting

$$[\underline{e}^{\pm}(z)] = \left[\underline{B}^{\pm}(z)\right] e^{\pm ikz} \tag{6.88}$$

in (6.86) and neglecting the second–order derivatives, the solutions of the ODEs (6.86) are obtained as

$$\begin{bmatrix} [\underline{e}^+(z)] \\ [\underline{e}^-(z)] \end{bmatrix} = \left[\underline{\underline{W}}^+(z - z_0)\right]\begin{bmatrix} [\underline{e}^+(z_0)] \\ [\underline{e}^-(z_0)] \end{bmatrix}, \qquad 0^+ \le z_0, z \le d/2^-, \tag{6.89}$$

and

$$\begin{bmatrix} [\underline{e}^+(z)] \\ [\underline{e}^-(z)] \end{bmatrix} = \left[\underline{\underline{W}}^-(z - z_0)\right]^{\mathsf{T}}\begin{bmatrix} [\underline{e}^+(z_0)] \\ [\underline{e}^-(z_0)] \end{bmatrix}, \qquad d/2^+ \le z_0, z \le d^-. \tag{6.90}$$

The transfer 4×4 matrices $\left[\underline{\underline{W}}^{\pm}(z)\right]$ are defined as

$$\left[\underline{\underline{W}}^{\pm}(z)\right] = \begin{bmatrix} P_-(z) & 0 & 0 & \pm Q_-(z) \\ 0 & P_+(z) & \pm Q_+(z) & 0 \\ 0 & \pm Q_+^*(z) & P_+^*(z) & 0 \\ \pm Q_-^*(z) & 0 & 0 & P_-^*(z) \end{bmatrix}, \qquad 0 \le z \le d/2, \tag{6.91}$$

where

$$P_{\pm}(z) = e^{\pm ih\pi z/\Omega}\left[\cosh(\Delta_{\mp}z) + \frac{i(k \mp h\pi/\Omega)}{\Delta_{\mp}}\sinh(\Delta_{\mp}z)\right], \tag{6.92}$$

$$Q_{\pm}(z) = e^{\pm ih\pi z/\Omega}\left[\frac{ik_\delta}{\Delta_{\mp}}\sinh(\Delta_{\mp}z)\right], \tag{6.93}$$

$k_\delta = k_0\delta_n$, $\delta_n = \delta_\epsilon/2\bar{n}$, and $\Delta_{\pm} = \sqrt{k_\delta^2 - (k \pm h\pi/\Omega)^2}$.

The structure of $\left[\underline{\underline{W}}^{\pm}(z)\right]$ confirms the decoupling of LCP and RCP phasors for all $z \in (0, d)$, within the approximate framework of coupled–wave theory. Furthermore, it is worth mentioning that neglect of the terms containing $e^{\pm 2i(k+\pi/\Omega)z}$ in (6.86) would mean that

$$\left. \begin{array}{l} P_+(z)\delta_{h,-1} + P_-(z)\delta_{h,1} \Rightarrow e^{ikz} \\ Q_+(z)\delta_{h,-1} + Q_-(z)\delta_{h,1} \Rightarrow 0 \end{array} \right\}, \tag{6.94}$$

thereby defeating the purpose of explaining the crossover phenomenon in the chiral STF. Here, $\delta_{m,m'}$ is the Kronecker delta.

6.4.1.3 Solution of Boundary Value Problem

Expressions for $[\underline{E}_\perp(z)]$ and $[\underline{H}'_\perp(z)]$ are to be obtained after combining (6.89) and (6.90) with the boundary conditions on the interfaces $z = 0$, $z = d/2$, and $z = d$. The continuity of the tangential components of the electric and magnetic field phasors across these interfaces implies that $[\underline{E}_\perp(z)]$ and $[\underline{H}'_T(z)] = (ik_0)^{-1}\frac{d}{dz}[\underline{E}_\perp(z)]$ should be continuous, according to the ODEs (6.76).

Consistently with the preceding developments, it is useful to define the circular counterparts of $[\underline{E}_\perp(z)]$ and $[\underline{H}'_T(z)]$ as

$$[\underline{\tilde{E}}_\perp(z)] = [\underline{\underline{Y}}]^\dagger [\underline{E}_\perp(z)] = [\underline{\tilde{e}}^+(z)] + [\underline{\tilde{e}}^-(z)] \tag{6.95}$$

and

$$[\underline{\tilde{H}}'_T(z)] = [\underline{\underline{Y}}]^\dagger [\underline{H}'_T(z)] = (ik_0)^{-1}\left[\frac{d}{dz}[\underline{\tilde{e}}^+(z)] + \frac{d}{dz}[\underline{\tilde{e}}^-(z)]\right], \tag{6.96}$$

where the superscript $\dagger$ denotes the Hermitian adjoint [66]. These 2–column vectors must be continuous across the interfaces as well, i.e.,

$$\left.\begin{array}{l}[\underline{\tilde{E}}_\perp(z^-)] = [\underline{\tilde{E}}_\perp(z^+)] \\ [\underline{\tilde{H}}'_T(z^-)] = [\underline{\tilde{H}}'_T(z^+)]\end{array}\right\}, \quad z \in \{0, d/2, d\}. \tag{6.97}$$

In accordance with (6.89), (6.90), (6.95), and (6.96), one can obtain the expressions

$$\begin{bmatrix}[\underline{\tilde{E}}_\perp(z)] \\ [\underline{\tilde{H}}'_T(z)]\end{bmatrix} = [\underline{\underline{Z}}^+]\begin{bmatrix}[\underline{\tilde{e}}^+(z)] \\ [\underline{\tilde{e}}^-(z)]\end{bmatrix}, \quad z \in \{0^+, d/2^-\}, \tag{6.98}$$

and

$$\begin{bmatrix}[\underline{\tilde{E}}_\perp(z)] \\ [\underline{\tilde{H}}'_T(z)]\end{bmatrix} = [\underline{\underline{Z}}^-]\begin{bmatrix}[\underline{\tilde{e}}^+(z)] \\ [\underline{\tilde{e}}^-(z)]\end{bmatrix}, \quad z \in \{d/2^+, d^-\}, \tag{6.99}$$

where the 4×4 matrices

$$[\underline{\underline{Z}}^\pm] = [\underline{\underline{Z}}(\bar{n})] \mp \delta_n \begin{bmatrix} 0 & 0 & 0 & 0 \\ 0 & 0 & 0 & 0 \\ 0 & 1 & 0 & -1 \\ 1 & 0 & -1 & 0 \end{bmatrix}, \quad [\underline{\underline{Z}}(\sigma)] = \begin{bmatrix} [\underline{\underline{I}}]_2 & [\underline{\underline{I}}]_2 \\ \sigma[\underline{\underline{I}}]_2 & -\sigma[\underline{\underline{I}}]_2 \end{bmatrix}. \tag{6.100}$$

The values of the field phasors at $z = 0^-$ and $z = d^+$ should be related to the electromagnetic field in the two half–spaces. The incident, reflected, and transmitted plane waves propagate along the z axis in the two half–spaces $z \leq 0$ and $z \geq d$. Following Venugopal and Lakhtakia [27, 28], and in analogy with

(6.88) and (6.95), the boundary value problem is reduced to the solution of the algebraic equation

$$\left[\begin{array}{c}[\mathcal{T}]\\ {[0]_2}\end{array}\right]=\left[\underline{\underline{\tau}}\right]\left[\begin{array}{c}[\mathcal{A}]\\ {[\mathcal{R}]}\end{array}\right], \qquad (6.101)$$

where $[\mathcal{A}]$, $[\mathcal{R}]$, and $[\mathcal{T}]$, respectively, comprise amplitudes of the LCP and RCP components of the incident, reflected, and transmitted plane waves, and the 4×4 transmission matrix

$$\left[\underline{\underline{\tau}}\right]=\left[\underline{\underline{Z}}_{hs}\right]^{-1}\left[\underline{\underline{Z}}^{-}\right]\left[\underline{\underline{\hat{\tau}}}\right]\left[\underline{\underline{Z}}^{+}\right]^{-1}\left[\underline{\underline{Z}}_{hs}\right] \qquad (6.102)$$

employs the matrix

$$\left[\underline{\underline{\hat{\tau}}}\right]=\left[\underline{\underline{W}}^{-}(d/2)\right]\left[\underline{\underline{Z}}^{-}\right]^{-1}\left[\underline{\underline{Z}}^{+}\right]\left[\underline{\underline{W}}^{+}(d/2)\right]. \qquad (6.103)$$

The transmission matrix $\left[\underline{\underline{\hat{\tau}}}\right]$ of (6.103), though specified here for the central $90°$ twist defect, applies in a general sense for any ϕ.

Although the transmission matrix $\left[\underline{\underline{\tau}}\right]$ seems very complicated, its replacement by $\left[\underline{\underline{\hat{\tau}}}\right]$ in (6.101) is found to generate values of $[\mathcal{R}]$ and $[\mathcal{T}]$ that exhibit the main spectral features of the circular Bragg phenomenon as well as those due to the central twist defect. In physical terms, the replacement of $\left[\underline{\underline{\tau}}\right]$ by $\left[\underline{\underline{\hat{\tau}}}\right]$ for obtaining $[\mathcal{R}]$ and $[\mathcal{T}]$ amounts to ignoring the index mismatch across the interfaces $z = 0$ and $z = d$. Indeed, when $|\delta_n| << \bar{n}$ and $n_{hs} = \bar{n}$, the chiral STF can be said to be index–matched to the medium in the two half–spaces, and $\left[\underline{\underline{\tau}}\right] \simeq \left[\underline{\underline{\hat{\tau}}}\right]$ would then be very true. Furthermore, it was ascertained from numerous computations that substantial development of circular Bragg phenomenon requires

$$N_d \geq |b|^{-1}, \qquad (6.104)$$

where $b = \delta_n/\bar{n}$ is the relative local birefringence.

6.4.2 Genesis of Crossover Phenomenon

Now, the optical response of a chiral STF with a central $90°$ twist defect is considered. Figures 6.11 and 6.12 show the spectrums of the co–polarized reflectances and transmittances obtained from the solution of (6.101) when $N_d = 100$ and $N_d = 600$, respectively. As expected, the coupled–wave theory captures the already–known phenomena that

- a co–handed reflectance hole, accompanied by a co–handed transmittance peak, occurs in the center of the Bragg regime $\lambda^{Br}_{0_{CWT}} = 2\bar{n}\Omega$ for relatively small N_d (Figure 6.11), disappearing for large N_d; and

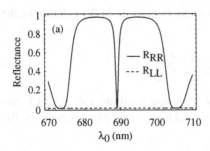
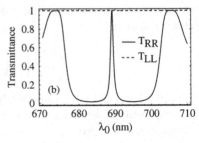

FIGURE 6.11. (a) Reflectances R_{LL} and R_{RR} and (b) transmittances T_{LL} and T_{RR}, computed for a structurally right–handed chiral STF with a central 90°–twist defect. The following parameters were used for coupled–wave theory calculations: $\epsilon_c = 1.7029^2$, $\epsilon_d = 1.7429^2$, $n_r = \bar{n} = 1.7230$, $h = 1$, $\Omega = 200$ nm, and $d = 100\Omega$. A co–handed reflectance hole and a co–handed transmittance peak must be noted.

- a cross–handed transmittance hole, accompanied by a cross–handed reflectance peak, occurs in the center of the Bragg regime for large N_d (Figure 6.12).

Typically, the bandwidth of the cross–handed transmittance hole is extraordinarily small, and is independent of N_d beyond its crossover value N_d^{co}.

In order to analyze the consequences of introducing the central 90° twist defect, recourse to the matrix $\left[\hat{\underline{\underline{\tau}}}\right]$ must be taken. This matrix can be decomposed as

$$\left[\hat{\underline{\underline{\tau}}}\right] = \left[\underline{\underline{W}}^-(d/2)\right]\left[\underline{\underline{W}}^+(d/2)\right] + \left[\underline{\underline{W}}^-(d/2)\right]\left[\underline{\underline{\Theta}}(b)\right]\left[\underline{\underline{W}}^+(d/2)\right], \quad (6.105)$$

where all nonzero entries of the 4×4 matrix

$$\left[\underline{\underline{\Theta}}(b)\right] = \left[\underline{\underline{Z}}^-\right]^{-1}\left[\underline{\underline{Z}}^+\right] - \left[\underline{\underline{I}}\right]_4, \quad (6.106)$$

are of $\mathcal{O}(b)$. This decomposition is not arbitrary but has physical meaning. The first term on the right side of (6.105), all by itself, would give rise to total

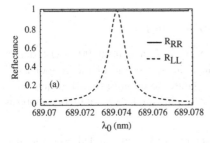
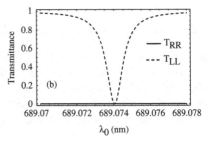

FIGURE 6.12. Same as Figure 6.11, but for $d = 600\Omega$. A cross–handed transmittance hole and a cross–handed reflectance peak must be noted.

transmission in the central part of the Bragg regime. In contrast, were $\left[\underline{\hat{t}}\right] = \left[\underline{\underline{W}}^-(d/2)\right]\left[\underline{\underline{\Theta}}(b)\right]\left[\underline{\underline{W}}^+(d/2)\right]$, total reflection would occur in the entire Bragg regime, for all $N_d \geq |b|^{-1}$. Thus, the second term on the right side of (6.105) describes a CP mirror.

As $|b| << 1$, the second term on the right side of (6.105) may be viewed as perturbing the leading term $\left[\underline{\underline{W}}^-(d/2)\right]\left[\underline{\underline{W}}^+(d/2)\right]$. Even so, the former is essential to the elucidation of the crossover phenomenon. For convenience, the weight of the second term in relation to the first term is defined

$$\alpha_\infty = \frac{\left\| \left[\underline{\underline{W}}^-(d/2)\right]\left[\underline{\underline{\Theta}}(b)\right]\left[\underline{\underline{W}}^+(d/2)\right] \right\|_\infty}{\left\| \left[\underline{\underline{W}}^-(d/2)\right]\left[\underline{\underline{W}}^+(d/2)\right] \right\|_\infty}, \qquad (6.107)$$

where $||.||$ is the L_∞–norm of matrix [66]. According to (6.91)–(6.93),

$$\alpha_\infty \approx \begin{cases} |b|, & \text{for } |b\sinh(\pi b N_d)| << 1, \\ |b/\beta|, & \text{for } |b\sinh(\pi b N_d)| >> 1, \end{cases} \qquad (6.108)$$

where $\beta = (k - \pi/\Omega)/k_\delta$. Clearly, α_∞ is independent of N_d in either of the N_d ranges specified in (6.108); therefore, both terms on the right side of (6.105) are similar to each other in the sense of L_∞ norms in those two N_d ranges. Furthermore, a wavelength regime can be mapped to a β regime uniquely. When $\beta = 0$, $\lambda_0 = \lambda_{0_{CWT}}^{Br}$; therefore, the neighborhood of $\lambda_{0_{CWT}}^{Br}$ can be equivalently specified through the β neighborhood of 0.

When N_d is relatively small (but larger than $|b|^{-1}$), $\left\| \left[\underline{\underline{W}}^\pm(d/2)\right] \right\|_\infty = \mathcal{O}(1)$ and $\alpha_\infty \approx |b| << 1$; thus, the second term on the right side of (6.105) can be ignored in favor of the first term, so that $\left[\underline{\hat{t}}\right] \approx \left[\underline{\underline{W}}^-(d/2)\right]\left[\underline{\underline{W}}^+(d/2)\right]$. According to (6.92) and (6.93), $\left[\underline{\underline{W}}^-(d/2)\right]\left[\underline{\underline{W}}^+(d/2)\right]$ becomes almost equal to $\left[\underline{\underline{I}}\right]_4$ for $\lambda_0 \sim \lambda_{0_{CWT}}^{Br}$, which gives rises to almost total transmission in the center of Bragg regime, as confirmed by Figure 6.11b for $N_d = 100$. In other words, the chiral STF with the central 90° twist defect becomes optically transparent in a small neighborhood of $\lambda_{0_{CWT}}^{Br}$, for both LCP and RCP plane waves. Another interpretation is that the phase difference between the field phasors at $z = 0^+$ and $z = d/2^-$ is exactly the opposite of the phase difference between the field phasors at $z = d/2^+$ and $z = d^-$.

When N_d is significantly large, the foregoing picture changes completely. No longer can $\left[\underline{\hat{t}}\right]$ be considered approximately equal to $\left[\underline{\underline{W}}^-(d/2)\right]\left[\underline{\underline{W}}^+(d/2)\right]$, because $\alpha_\infty \approx |b/\beta| \to +\infty$ as $\lambda_0 \to \lambda_{0_{CWT}}^{Br}$. The first term on the right side of (6.105) is still almost equal to $\left[\underline{\underline{I}}\right]_4$ at $\lambda_0 = \lambda_{0_{CWT}}^{Br}$ — and therefore indicates total transmission all by itself in the central part of the Bragg regime — although the bandwidth of that feature falls exponentially as N_d increases. But, the second term interferes with the first term in the neighborhood of $\lambda_{0_{CWT}}^{Br}$ so that $\left[\underline{\hat{t}}\right]$ is isomorphic to the transmittance matrix of a defect–free chiral STF — except in a tiny neighborhood of $\beta = b$ wherein $\alpha_\infty \approx 1$. *It must be the L_∞–norm–equivalence of the two terms that engenders the total–reflection feature in the tiny neighborhood of $\beta = b$*, as confirmed by Figure 6.12a for $N_d = 600$. This conclusion applies to any arbitrary $\phi \neq m\pi$, $(m \in \mathbb{Z})$, as proved by further investigations.

Finally, although the foregoing analysis successfully provides a rigorous explanation of the crossover phenomenon for either small or significantly large values of N_d, it is not able to yield an estimate of N_d^{co}. The reason for that incapability is the neglect of the second–order derivatives in the coupled–wave ODEs (6.86).

6.5 Concluding Remarks

In this chapter, slanted chiral STFs were optically investigated to couple the characteristic optical responses of volume and diffraction gratings in thin films. A robust numerical solution procedure was devised in Section 6.2 for computation of planewave response of slanted chiral STFs. The same solution procedure was employed to analyze the wave–resonance phenomenon of a slanted chiral STF with a central twist defect in Section 6.3. A remarkable crossover phenomenon was found in slanted chiral STFs for the spectral holes initiated by the twist defect. The genesis of the crossover phenomenon in chiral STFs was mathematically elucidated in Section 6.4, by an application of the coupled–wave theory. The conclusions drawn from the studies discussed in the preceding sections are summarized as follows:

- The circular Bragg phenomenon becomes partly nonspecular when the slant angle $\alpha \neq 0$, such that the co–handed Bragg reflection occurs in the Floquet harmonic of order $n = \pm 2$ for $\alpha \gtrless 0$. This characteristic can be harnessed to design slanted chiral STFs as circular–polarization beamsplitters and couplers.
- The Bragg regime blue–shifts as $|\alpha|$ increases, and the width of the Bragg regime decreases to zero due to the influence of a Rayleigh–Wood anomaly of order $n = \pm 2$ for $\alpha \gtrless 0$.
- The angular spread of the Bragg regime is significantly asymmetric because of the subversion of the circular Bragg phenomenon by the Rayleigh–Wood anomaly.
- In the presence of a central twist defect, spectral holes emerge within the Bragg regime, thereby indicating the occurrence of wave resonance therein. Two types of spectral holes are excitable by circularly polarized plane waves — one is the co–handed reflectance hole, and the second is the cross–handed transmittance hole.
- There is a crossover phenomenon associated with the evolution of the two types of spectral holes with the increase of thickness of the chiral STF: When the thickness is small, a co–handed reflectance hole occurs in the Bragg regime; as the thickness increases, the co–handed reflectance

hole diminishes to vanish eventually, and is replaced by a cross–handed transmittance hole.

- The bandwidth of the cross–handed transmittance hole is significantly smaller than that of the co–handed reflectance hole.
- The co–handed reflectance hole is nonspecular for $\alpha \neq 0$, while the cross–handed transmittance hole is always specular.
- An approximate but closed–form solution for axial wave propagation in a chiral STF with a central 90° twist defect is obtained in terms of a coupled–wave–theory 4×4 transmission matrix.
- The genesis of the crossover phenomenon in the chiral STF is mathematically elucidated by the spectral characteristics of the transmission matrix: As it can be decomposed into two terms, the first term favors total transmission in the middle of the Bragg regime, while the second term favors total reflection in the whole Bragg regime. When the thickness of the chiral STF is relatively small, the first term dominates and gives rise to a co–handed reflectance hole in the center of the Bragg regime. As the thickness of the chiral STF increases, the second term becomes significant and interferes with the first term such that the transmission matrix is isomorphic to that of a defect–free chiral STF — except in a tiny wavelength–regime wherein the two terms become identical to each other in the L_∞–norm sense to engender the total–reflection feature. Hence, the co–handed reflectance hole diminishes to vanish eventually as the thickness increases, and is replaced by a cross–handed transmittance hole.

Acknowledgments. The author is grateful to Professor A. Lakhtakia (Pennsylvania State University) for his initiative efforts and long–term support in the investigation of slanted chiral STFs. The author is appreciative of Professor R. Messier and Professor M. W. Horn (Pennsylvania State University) for their helpful discussions and collaborations on relevant topics in past years. Special thanks are given to William A. Stanton (Micron Technology, Inc, USA) for his generosity in providing this author with the pivotal chance to observe real–time evolution of nanotechnology.

References

1. A. Lakhtakia and R. Messier, *Sculptured Thin Films: Nanoengineered Morphology and Optics*, SPIE Press, Bellingham, WA (2005).
2. H. A. Macleod, *Thin–Film Optical Filters*, Institute of Physics, Bristol, UK (2001).
3. I. J. Hodgkinson and Q. h. Wu, *Birefringent Thin Films and Polarizing Elements*, World Scientific, Singapore (1997).
4. A. Lakhtakia, Sculptured thin films: accomplishments and emerging uses, *Mater. Sci. Eng. C* **19**, 427–434 (2001).
5. I. J. Hodgkinson and Q. h. Wu, Inorganic chiral optical materials, *Adv. Mater.* **13**, 889–897 (2001).

6. A. Lakhtakia and R. Messier, The past, the present, and the future of sculptured thin films, in *Introduction to Complex Mediums for Optics and Electromagnetics*, edited by W. S. Weiglhofer and A. Lakhtakia, SPIE Press, Bellingham, WA (2003).

7. A. Lakhtakia and R. Messier, Sculptured thin films—I. Concepts, *Mater. Res. Innovat.* **1**, 145–148 (1997).

8. R. Messier and A. Lakhtakia, Sculptured thin films—II. Experiments and applications, *Mater. Res. Innovat.* **2**, 217–222 (1999).

9. V. C. Venugopal and A. Lakhtakia, Sculptured thin films: Conception, optical properties, and applications, in *Electromagnetic Fields in Unconventional Materials and Structures*, Edited by N. Singh and A. Lakhtakia, Wiley, New York, pp. 151–216 (2000).

10. M. Suzuki and Y. Taga, Integrated sculptured thin films, *Jpn. J. Appl. Phys. Pt. 2* **40**, L358–L359 (2001).

11. M. W. Horn, M. D. Pickett, R. Messier, and A. Lakhtakia, Blending of nanoscale and microscale in uniform large–area sculptured thin–film architectures, *Nanotechnology* **15**, 303–310 (2004).

12. J. A. Thornton, High rate thick film growth, *Annu. Rev. Mater. Sci.* **7**, 239–260 (1977).

13. V. C. Venugopal and A. Lakhtakia, Low–permittivity nanocomposite materials using sculptured thin film technology, *J. Vac. Sci. Technol. B* **18**, 32–36 (2000).

14. A. Lakhtakia, On determining gas concentration using thin–film helicoidal bianisotropic medium bilayers, *Sens. Actuat. B* **52**, 243–250 (1998).

15. A. Lakhtakia, M. W. McCall, J. A. Sherwin, Q. H. Wu, and I. J. Hodgkinson, Sculptured–thin–film spectral holes for optical sensing of fluids, *Opt. Commun.* **194**, 33–46 (2001).

16. J. A. Sherwin and A. Lakhtakia, Nominal model for structure–property relations of chiral dielectric sculptured thin films, *Math. Comput. Model.* **34**, 1499–1514 (2001); corrections: **35**, 1355–1363 (2002).

17. J. A. Sherwin and A. Lakhtakia, Nominal model for the optical response of a chiral sculptured thin film infiltrated with an isotropic chiral fluid, *Opt. Commun.* **214**, 231–245 (2002).

18. J. A. Sherwin, A. Lakhtakia, and I. J. Hodgkinson, On calibration of a nominal structure–property relationship model for chiral sculptured thin films by axial transmittance measurements, *Opt. Commun.* **209**, 369–375 (2002).

19. J. A. Sherwin and A. Lakhtakia, Nominal model for the optical response of a chiral sculptured thin film infiltrated by an isotropic chiral fluid—oblique incidence, *Opt. Commun.* **222**, 305–329 (2003).

20. R. Messier, T. Gehrke, C. Frankel, V. C. Venugopal, W. Otaño, and A. Lakhtakia, Engineered sculptured nematic thin films, *J. Vac. Sci. Technol. A* **15**, 2148–2152 (1997).

21. N. O. Young and J. Kowal, Optically active fluorite films, *Nature* **183**, 104–105 (1959).

22. K. Robbie, M. J. Brett, and A. Lakhtakia, First thin film realization of a helicoidal bianisotropic medium, *J. Vac. Sci. Technol. A* **13**, 2991–2993 (1995).

23. M. D. Pickett and A. Lakhtakia, On gyrotropic chiral sculptured thin films for magneto–optics, *Optik* **113**, 367–371 (2002).

24. A. Lakhtakia, On the genesis of Post constraint in modern electromagnetism, *Optik* **115**, 151–158 (2004).

25. C. M. Krowne, Electromagnetic theorems for complex anisotropic media, *IEEE Trans. Antennas Propagat.* **32**, 1224–1230 (1984).

26. A. Lakhtakia (Ed.), *Selected Papers on Linear Optical Composite Materials*, SPIE Press, Bellingham, WA (1996).

27. V. C. Venugopal and A. Lakhtakia, Electromagnetic plane–wave response characteristics of non–axially excited slabs of dielectric thin–film helicoidal bianisotropic mediums, *Proc. R. Soc. Lond on A* **456**, 125–161 (2000).

28. V. C. Venugopal and A. Lakhtakia, On absorption by non–axially excited slabs of dielectric thin–film bianisotropic mediums, *Eur. J. Phys. Appl. Phys.* **10**, 173–184 (2000).

29. M. D. Pickett, A. Lakhtakia, and J. A. Polo, Spectral responses of gyrotropic chiral sculptured thin films to obliquely incident plane waves, *Optik* **115**, 393–398 (2004).

30. M. W. McCall and A. Lakhtakia, Development and assessment of coupled wave theory of axial propagation in thin–film helicoidal bianisotropic media. Part 1: reflectances and transmittances, *J. Mod. Opt.* **47**, 973–991 (2000); corrections: **50**, 2807 (2003).

31. M. W. McCall, Axial electromagnetic wave propagation in inhomogeneous dielectrics, *Math. Comput. Model.* **34**, 1483–1497 (2001).

32. M. W. McCall and A. Lakhtakia, Explicit expressions for spectral remittances of axially excited chiral sculptured thin films, *J. Mod. Opt.* **51**, 111–127 (2004).

33. Q. Wu, I. J. Hodgkinson, and A. Lakhtakia, Circular polarization filters made of chiral sculptured thin films: Experiments and simulation results, *Opt. Eng.* **39**, 1863–1868 (2000).

34. A. Lakhtakia, Dielectric sculptured thin films for polarization–discriminatory handedness–inversion of circularly polarized light, *Opt. Eng.* **38**, 1596–1602 (1999).

35. I. J. Hodgkinson, A. Lakhtakia, and Q. h. Wu, Experimental realization of sculptured–thin–film polarization–discriminatory light–handedness inverters, *Opt. Eng.* **39**, 2831–2834 (2000).

36. I. J. Hodgkinson, Q. h. Wu, A. Lakhtakia, and M. W. McCall, Spectral–hole filter fabricated using sculptured thin–film technology, *Opt. Commun.* **177**, 79–84 (2000).

37. I. J. Hodgkinson, Q. h. Wu, K. E. Thorn, A. Lakhtakia, and M. W. McCall, Spacerless circular–polarization spectral–hole filters using chiral sculptured thin films: Theory and experiment, it Opt. Commun. **184**, 57–66 (2000).

38. F. Wang, A. Lakhtakia, and R. Messier, On piezoelectric control of the optical response of sculptured thin films, *J. Mod. Opt.* **50**, 239–249 (2003).

39. F. Wang, A. Lakhtakia, and R. Messier, Towards piezoelectrically tunable chiral sculptured thin film lasers, *Sens. Actuat. A* **102**, 31–35 (2002).

40. J. B. Geddes III and A. Lakhtakia, Reflection and transmission of optical narrow–extent pulses by axially excited chiral sculptured thin films, *Eur. Phys. J. Appl. Phys.* **13**, 3–14 (2001); corrections: **16**, 247 (2001).

41. J. B. Geddes III and A. Lakhtakia, Pulse–coded information transmission across an axially excited chiral sculptured thin film in the Bragg regime, *Microwave Opt. Technol. Lett.* **28**, 59–62 (2001).

42. A. T. Wu, M. Seto, and M. J. Brett, Capacitive SiO humidity sensors with novel microstructures, *Sens. Mater.* **11**, 493–505 (2000).

43. E. E. Steltz and A. Lakhtakia, Theory of second–harmonic–generated radiation from chiral sculptured thin films for bio–sensing, *Opt. Commun.* **216**, 139–150 (2003).

44. J. M. Jarem and P. P. Banerjee, *Computational Methods for Electromagnetic and Optical Systems*, Dekker, New York (2000).

45. D. Maystre (Ed.), *Selected Papers on Diffraction Gratings*, SPIE Press, Bellingham, WA (1993).
46. A. Lakhtakia, V. K. Varadan, and V. V. Varadan, Scattering by periodic achiral–chiral interfaces, *J. Opt. Soc. Am. A* **6**, 1675–1681 (1989); corrections: **7**, 951 (1990).
47. F. Wang and A. Lakhtakia (Eds.), *Selected Papers on Nanotechnology- Theory and Modeling*, SPIE Press, Bellingham, WA (2006).
48. R. Messier, V. C. Venugopal, and P. D. Sunal, Origin and evolution of sculptured thin films, *J. Vac. Sci. Technol. B* **18**, 1538–1545 (2000).
49. R. Messier, A. Lakhtakia, V. C. Venugopal, and P. D. Sunal, Sculptured thin films: Engineered nanostructural materials, *Vac. Technol. Coat.* **2**(10), 40–47 (October 2001).
50. F. Wang, A. Lakhtakia, and R. Messier, Coupling of Raleigh–Wood anomalies with the circular Bragg phenomenon in the slanted sculptured thin films, *Eur. Phys. J. Appl. Phys.* **20**, 91–103 (2002); corrections: **24**, 91 (2003).
51. F. Wang and A. Lakhtakia, Lateral shifts of optical beams on reflection by slanted chiral sculptured thin films, *Opt. Commun.* **235**, 107–132 (2004).
52. F. Wang and A. Lakhtakia, Response of slanted chiral sculptured thin films to dipolar sources, *Opt. Commun.* **235**, 133–151 (2004).
53. F. Wang and A. Lakhtakia, Specular and nonspecular, thickness–dependent, spectral holes in a slanted chiral sculptured thin film with a central twist defect, *Opt. Commun.* **215**, 79–92 (2003).
54. F. Wang and A. Lakhtakia, Third method for generation of spectral holes in chiral sculptured thin films, *Opt. Commun.* **250**, 105–110 (2005).
55. H. A. Haus and C. V. Shank, Antisymmetric taper of distributed feedback lasers, *IEEE J. Quantum Electron.* **12**, 532–539 (1974).
56. E. Yablonovitch, Inhibited spontaneous emission in solid–state physics and electronics, *Phys. Rev. Lett.* **58**, 2059–2062 (1987).
57. J. Schmidtke, W. Stille, and H. Finkelmann, Defect mode emission of a dye doped cholesteric polymer network, *Phys. Rev. Lett.* **90**, 083902 (2003).
58. I. J. Hodgkinson, Q. h. Wu, L. De Silva, M. Arnold, M. W. McCall, and A. Lakhtakia, Supermodes of chiral photonic filters with combined twist and layer defects, *Phys. Rev. Lett.* **91**, 223903 (2003).
59. V. I. Kopp and A. Z. Genack, Twist defect in chiral photonic structures, *Phys. Rev. Lett.* **89**, 033901 (2002).
60. M. Becchi, S. Ponti, J. A. Reyes, and C. Oldano, Defect mode in helical photonic crystals: An analytic approach, *Phys. Rev. B* **70**, 033103 (2004).
61. J. Schmidtke and W. Stille, Photonic defect modes in cholesteric liquid crystal films, *Eur. Phys. J. E* **12**, 553–564 (2003).
62. F. Wang and A. Lakhtakia, Optical crossover phenomenon due to a central 90°–twist defect in a chiral sculptured thin film and chiral liquid crystal, *Proc. R. Soc. London Ser. A* **461**, 2985–3004 (2005).
63. F. Wang and A. Lakhtakia, Defect modes in multisection helical photonic crystals, *Opt. Express* **13**, 7319–7335 (2005).
64. C. Kittel, *Introduction to Solid State Physics*, 4th ed., Wiley Eastern, New Delhi, Chapter 13 (1974).
65. A. Lakhtakia, Spectral signatures of axially excited slabs of dielectric thin-film helicoidal bianisotropic mediums, Eur. Phys. J. Appl. Phys. **8**, 129–137 (1999).
66. H. Lütkepohl, *Handbook of Matrices*, Wiley, New York (1996).

67. N. Chateau and J. Hugonin, Algorithm for the rigorous coupled–wave analysis of grating diffraction, *J. Opt. Soc. Am. A* **11**, 1321–1331 (1994).
68. L. Li, Multilayer modal method diffraction gratings of arbitrary profile, depth, and permittivity, *J. Opt. Soc. Am. A* **10**, 2581–2591 (1993).
69. M. G. Moharam, E. B. Grann, D. A. Pommet, and T.K. Gaylord, Formulation for stable and efficient implementation of the rigorous coupled–wave analysis of binary gratings, *J. Opt. Soc. Am. A* **12**, 1068–1076 (1995).
70. M. G. Moharam, D. A. Pommet, E. B. Grann, and T. K. Gaylord, Stable implementation of the rigorous coupled–wave analysis for surface–relief gratings: Enhanced transmittance matrix approach, *J. Opt. Soc. Am. A* **12**, 1077–1086 (1995).
71. I. J. Hodgkinson, Q. h. Wu, B. Knight, A. Lakhtakia, and K. Robbie, Vacuum deposition of chiral sculptured thin films with high optical activity, *Appl. Opt.* **39**, 642–649 (2000).
72. P. Kipfer, R. Klappert, H. P. Herzig, J. Grupp, and R. Dändliker, Improved red color with cholesteric liquid crystals in Bragg reflection mode, *Opt. Eng.* **41**, 638–646 (2002).
73. F. Wang, M. W. Horn, and A. Lakhtakia, Rigorous electromagnetic modeling of near–field phase–shifting contact lithography, *Microelectron. Eng.* **71**, 34–53 (2004).
74. C. Oldano, Comment on "Twist defect in chiral photonic structures", *Phys. Rev. Lett.* **91**, 259401 (2003).
75. S. F. Nagle and A. Lakhtakia, Attenuation and handedness of axial propagation modes in a cholesteric liquid crystal, *Microwave Opt. Technol. Lett.* **7**, 749–752 (1994).
76. V. A. Yakubovich and V. M. Starzhinskii, *Linear Differential Equations with Periodic Coefficients*, Wiley, New York (1975).

Erratum

© Springer Science + Business Media, Inc. 2007
Publisher's Erratum to:
Frontiers in Surface Nanophotonics:
Principles and Applications
David L. Andrews and Zeno Gaburro (Editors), 2007

The publisher inserted an incorrect figure as Fig. 1.2 on page 7.
The correct figure is given below.

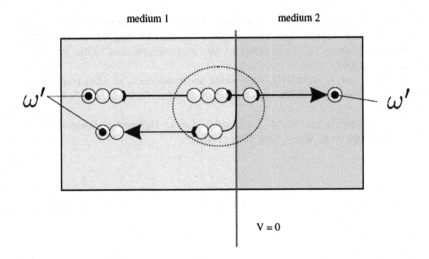

Page 4. Line below equation (1.10):
"The solutions of Equations (1.2) are plane waves..." **should read**
"The solutions of Equations (1.9) are plane waves..."

Page 4. Line below equation (1.11):
"In the frame at rest with the medium, Equations (1.2)..." **should read**
"In the frame at rest with the medium, Equations (1.9)..."

Page 6. Line below Equation (1.19):
"as obtained by substituting Equations (3.1)..." **should read**
"as obtained by substituting Equations (1.17)..."

Index

Springer Series in
OPTICAL SCIENCES